The Appreciation of Birds

We see distinctly only what we know thoroughly.
—EDWARD A. WILSON

LOUIS J. HALLE

The Appreciation of Birds

✦

with drawings by Jens Gregersen

THE JOHNS HOPKINS UNIVERSITY PRESS   BALTIMORE AND LONDON

© 1989 The Johns Hopkins University Press
All rights reserved
Printed in the United States of America

The Johns Hopkins University Press, 701 West 40th Street,
Baltimore, Maryland 21211
The Johns Hopkins Press Ltd., London

The paper used in this publication meets the minimum requirements of American National Standard for Information Sciences—Permanence of Paper for Printed Library Materials, ANSI Z39.48-1984.

LIBRARY OF CONGRESS CATALOGING-IN-PUBLICATION DATA

Halle, Louis Joseph, 1910–
The appreciation of birds / Louis J. Halle.
p.   cm.
ISBN 0-8018-3869-X
1. Birds.   I. Title.
QL676.H232   1989
598—DC19   89-2347
CIP

To Herbert S. Bailey, Jr.,
in grateful appreciation of his moral and intellectual support
over many years.

FOREWORD

Natural history has an advantage over many of the other disciplines of natural science in that it can be immediately understood. No great technical knowledge is necessary to comprehend the basic phenomena, as is the case, for instance, in molecular biology or in physics, where a familiarity with calculus and other forms of mathematics is prerequisite. The opposite is the case with natural history; anyone can become beguiled. A parallel could be drawn to many other disciplines, especially in the arts, where an appreciation of music, literature, or the fine arts, or an understanding of history, can be enjoyed by both the naive beginner and the most sophisticated professional. It is undoubtedly the case that what the neophyte sees is quite different from what the expert sees, but in their own ways their enjoyment can be equal. I might even say (although I may be stretching the analogy too far) that the same is true of religion: the most uneducated and illiterate farm hand can find solace and joy in his belief to no less a degree than an archbishop or a professor of theology.

The ability to find inspiration and pleasure in nature comes first and foremost from studying nature itself. It happens on those walks in the woods or over the fields, where each plant, each insect, each vertebrate is something that might grasp our attention. It makes no difference that a person might only see the beetles, the butterflies, the flowers, or the birds, for they are all part of a wonderful setting that makes our hearts sing.

It may be at dawn, at dusk, or in the heat of midday, for there is an enchantment in just being out-of-doors that sets the right mood. This is even true for the hunter or the fisher, although their activities may seem far removed from natural history.

Of all the outdoor interests and pleasures of mankind, none has a greater following than that of watching birds. Their colors, their grace, and their fascinating behavior have captivated an enormous number of people. And, as it should be, the bird watcher may vary from the rawest enthusiastic beginner to the equally enthusiastic professional ornithologist. It is nature that starts us and keeps us going; that is the source and the fountain that endlessly supplies our interest with new things to see and understand.

The celebrated nineteenth-century zoologist Louis Agassiz is well known for his injunction to embryonic naturalists, "Study nature not books." He certainly did not follow his own advice; in fact, he committed the double sin of not only reading the books of others but also writing a number of his own. I would simply alter his aphorism and say, "Study nature *and* books." Reading is a different kind of passion, but in its own way equally important. On those dark winter evenings, what better way is there to expand our horizons, our understanding, and our knowledge? No end of "nature" television shows, however entertaining, can do what books do for us.

In this book, Louis Halle's first theme is the importance of melding book knowledge with field knowledge in the study of birds. What he does not say (and I will say it for him) is that this applies aptly to his own book, which holds many new insights and new knowledge of just the sort that will be particularly valuable to both the budding and the seasoned ornithologist. He shows how reading can heighten one's observations and one's sensitivity to the bird world around us; at the same time, the story he tells does this in a striking and absorbing way. For he is part of that guild of writer-naturalists who are masters of elegant and pleasing prose. His sentences shine in the tradition of Thoreau and W. H. Hudson.

There are a number of ways books can illuminate nature for us. They can provide detailed facts of the appearance and the habits of birds which enrich the observer, and this is the point Louis Halle makes in his opening passages. Another way is to make us think about more general problems and to ask larger questions. These have the dual function of helping our observations in the field and, as is done in this book, surreptitiously expanding our minds in such an engaging way that we hardly realize we have learned some important principles. My own bias is that the greatest achievement to which a naturalist can aspire is to find patterns in nature that give an inner meaning and a cohesion to a multitude of facts. Anyone can amass facts—great numbers of them—and this is all to the good; I do not mean to argue against it in any way. But intriguing generalizations and patterns that group those facts in such a way that they suddenly take on new life is to me an added and wonderful gift. That is precisely what Louis Halle has done in this book.

It is clear from his previous writings that he thinks about things; he worries about causes and consequences and tries to understand why things happen. He has asked profound questions about international politics (which is his profession) and about the cosmos, including physical, biological, and historical evolution (in his monumental *Out of Chaos*). In this engaging book he gently applies those analytic powers to sharpen our appreciation of birds. It is clear that in this he has been successful, and we are all in his debt.

JOHN TYLER BONNER

No one would deny that, to cultivate an appreciation of birds, one must observe them directly in the state of nature. However, if one limited oneself to such observation alone, the appreciation one acquired would amount to relatively little. For a full and true appreciation one must go to the books, even before observing in the field, because the understanding of what one sees depends on the knowledge one has already acquired.

The essential attributes of any species are far more than just those of the animated bundle of feathers that presents itself to the observer's eye at a particular time and place. What he sees, in itself, tells him nothing about the bird's life-history from birth to death, or about the way it supports itself or the way it reproduces itself, or about its ancestry and family relationships. For this he has to go to books; and, if appreciation is the objective, he had best go to them in advance of meeting the bird itself.

✧

The first thing one wants to know about any bird one observes is its identity, what species it belongs to. The unprepared observer, dependent on his own resources alone, notes what he can about its appearance in the time available (which may be short), so that he can later describe it to someone who will identify it for him by his description. Without prior knowledge from books, however, he doesn't even know what to look for. Perhaps what he notes is that it is small and gray, with dark streaks on its breast. But, when the person later consulted asks whether it had a thick or thin bill, or whether it had wing-bars, he has to confess that he had failed to note these points, not having known that the identification depended on them. If he had studied the books in advance, he would have known.

✧

If identification were all, there would be an alternative to the prior study of books. The novice could go out into the field with an experienced bird-watcher, who not only identified the various species for him but also taught him how to identify them for himself. Indeed, the best way to learn is to combine the study of books with just such instruction in the field. Such instruction is especially valuable for the identification of birds by their vocal performances, their calls and songs, because the written description of what comes to one's ears is bound to be less precise than the description of what one sees with one's eyes. (Whatever may be the case in the world of bats, the world in which we humans live is primarily visual; for our ears are not such fine instruments as our eyes.)

Listening to recordings of bird calls and songs can be helpful, but most of us are able to remember images (e.g., the plates in a field guide) with greater precision than we can irregular or nondescript sound-patterns. In any case, identification by sound generally requires that a bird give clear and extended utterance, and this one cannot count on. It is not by their voices that one normally identifies seabirds, hawks, herons, cranes, storks. Even the so-called songbirds, not all of which have songs, sing only for a brief season each year.

✧

It is true that the interest of many birders is limited to identification. They are less interested in getting to know the birds than in seeing how many species they can add to their "life-lists," or how many they can "get" in a day, or simply in "get-ting" rare ones. If their interest does not go beyond the playing of such games—and there is no law that says it should—they need not read books at all, or even observe for themselves. They need only go out on bird-walks led by persons who can do the identifying for them, confining themselves to adding each species name to the lengthening list of what they can then claim to have seen, if only in a glimpse too brief to allow ob-servation (or even to allow identification, for which they take the leader's word).

One appeal of such activity is social, the pleasure of spend-ing a day in the country with a group of good companions. This is legitimate in itself, but has nothing to do with the ap-preciation of birds, which is what concerns us here.

In any case, one learns for oneself only when one has to de-pend on one's own observation, the powers of which are de-veloped by practice. This is an argument for going into the field alone, at least occasionally. (In this increasingly crowded world, moreover, the enjoyment of solitude may, in itself, jus-tify this. One's thoughts, if one will go off alone with them, offer their own companionship.)

✧

Identification comes first. But whoever would appreciate birds cannot stop there. I had rather get to know the American robin intimately, becoming familiar with all its ways and how it lives, than to have had one glimpse of that rare bird, the blue petrel, that does not last long enough for anything more than identification. (I did have such a glimpse, lasting a second or two, but what appreciation of the blue petrel I have still comes entirely from books.)

Appreciation, in its degree, depends on knowledge, and the

largest part of knowledge, even of the American robin, is not accessible to the senses of the observer in the field.

I shall return to the robin.

✦

Up and down the river darts a blue-black swallow.

—A. A. MILNE

Picture a man who is newly arrived from Mars, standing by a river, seeing his first bird of any sort. An atmospheric haze, penetrated by the day's first sunlight, veils the trees on the op-posite shore, investing the scene with a golden glitter. Everything is motionless except a swallow, which darts this way and that, doubling on its track, rising and falling, with flicks and flickerings of its wings. In its continuous smooth move-ment, here and there and back again, it is silent music made visible. The bird wears beauty as the sky wears blue. No need of the information contained in books to appreciate such a sight.

What the man from Mars appreciates, however, is simply a performance in a landscape. Never even having heard of such a creature as a bird, he might suppose the performer to be a contrivance actuated by rubber-bands.

What, then, about the performer itself?

For an answer, one must go to the books.

✦

No one knows the bee who doesn't know the hive.

Like the bee, the swallow is not an existence in itself. It is not self-contained, nor is it confined to the particular here-and-now in which it is observed. It is inseparable from a so-ciety extensive in time and space. As the bee belongs to the hive, so the swallow (along with the bee) belongs to a web of life that extends worldwide, and reaches back over the billions of years in which it has been evolving.

✦

Identification entails the assignment of a name; but the vernacular names of birds, as opposed to the scientific, may not allow precision of identification, if only because they vary locally. So the species observed by the man from Mars, which occurs on both sides of the Atlantic, is called "barn swallow" in North America, where other species of what are there called "swallows" also occur (e.g., the "bank swallow"), but simply "swallow," without an adjective, in England, where other species of what Americans call "swallows" are called "martins." In pursuit of understanding, then, we resort to the scientific name, *Hirundo rustica,* which is accepted in all countries. Just as the double name, "John Brown," represents the individual and the family, which includes other individuals with other first names, so the scientific name of the swallow represents a like hierarchy—in this case, genus and species. There are, in the world, some twenty-five species of the genus *Hirundo,* *rustica* being only one—for example, *H. neoxena* in Australia, and *H. albigularis* in South Africa. *H. rustica* is by far the most abundant and widespread, breeding throughout the Northern Hemisphere (the far north excepted), and occurring during the northern winter throughout the continental masses of the Southern Hemisphere as well (Antarctica excepted). This cosmopolitanism enters into our appreciation. But we know about it only from books.

As the species *rustica* belongs to the genus *Hirundo,* so the genus *Hirundo* belongs to the swallow family, *Hirun-*

dinidæ, which includes some twenty other genera as well. The family *Hirundinidæ,* in turn, belongs to the order *Passeriformes* (the so-called perching birds, most familiar in field and garden), which belongs to the class *Aves,* which belongs to the subphylum *Vertebrata* of the phylum *Chordata,* which belongs to the *animal kingdom,* which belongs to the ultimate category, that of *life-on-earth. Life-on-earth* constitutes one web that relates the swallow to the human observer (*Homo sapiens,* of the family *Hominidæ,* of the order *Primates,* of the class *Mammalia,* of the subphylum *Vertebrata* . . . etc.). So the swallow, if not my brother, is at least a cousin, and all life is one.

When we study our own kind, in the science called anthropology, we study the stages of its evolution from primitive anthropoids; their evolution, in turn, from still more primitive mammals; and so on—all the way back to some microscopic ancestral cell of three or four billion years ago, the common ancestor of all life-on-earth. We may trace the swallow's evolution back, in the same fashion, to a common ancestor of man and swallow among the primitive vertebrates, beyond which we share one line of descent.

Should we not tell the man from Mars all this, so that he may have a proper appreciation of what he is seeing?

✦

The swallow has entered into the literature of twenty-five centuries, and this, too, heightens our appreciation.

Yea, the stork in the heaven knoweth her appointed times; and the turtle and the crane and the swallow observe the time of their coming. —JEREMIAH 8:7

All the summer long is the swallow a most instructive pattern of unwearied industry and affection; for, from morning

to night, while there is a family to be supported, she spends the whole day in skimming close to the ground, and exerting the most sudden turns and quick evolutions.
—GILBERT WHITE (SELBORNE, JANUARY 29, 1774)

Jeremiah knew that the swallows go south in winter, returning in spring. But Gilbert White, misinterpreting certain observations of his own, concluded that they passed the winter in hibernation beneath lakes and mill-ponds.

✦

And what should they know of England who only England know? —KIPLING

I thought I knew the American robin, *Turdus migratorius,* when that was the only robin I knew. I lived in North America, where it was the sole representative of the genus *Turdus.*

As I grew in knowledge, by reading, I found that there were other species of *Turdus* in Europe, and that *Turdus* was considered the typical genus of the thrush family, *Turdidæ.* But this tended to remain academic in the absence of field observation (which, to be sure, must accompany reading as reading must accompany such observation). The fieldfare, *Turdus pilaris,* the blackbird, *T. merula,* the redwing, *T. iliacus,* the song thrush, *T. philomelos,* and the mistle thrush, *T. viscivorus,* had no living reality for me. Then, however, in my young manhood, I found myself living and traveling in Central and South America, where I was surprised to  come upon American robins that were not quite American robins, birds obviously related to our robin but not identical—such as the rufous-collared robin, *T. rufitorques,* the white-throated robin, *T. assimilis,* the clay-colored robin, *T. grayi,* the mountain robin, *T. plebejus,* the black robin, *T. infuscatus,* the red-bellied robin, *T. rufiventris,* and the chestnut-bellied robin, *T. fuliventris.* Immediately, my appreciation of our own robin was enlarged, for I saw it as one in a distinctive series of similar thrushes (similar in calls and behavior, as well as appearance) extending from Alaska to Patagonia—an outlying member north of the Rio Grande. (Should we not, under the circumstances, give it the adjective "North American"?)

It was the same with the mockingbird, *Mimus polyglottis,* which I belatedly discovered to be only one in a series of mockingbirds, all confined to the New World and extending southward to the Straits of Magellan. My appreciation of the "North American" mockingbird, this outlying member of the series, was enlarged by acquaintance with siblings of which I had not previously been told.

Although the books might have told me all this, they didn't. In what book on the birds of North America, designed for the general reader, can one find an account of the American robin that begins by reporting that it is one of a series of closely similar large thrushes extending throughout the New World from Alaska to Patagonia?

✧

I suggest to the directors of natural-history museums that, in their exhibitions of mounted birds, they include whole series together—e.g., all the robins or all the mockingbirds *seriatim,* so that the viewer can appreciate genus as well as species.

Let zoos do the same with live birds in cages.

✧

It has tended to be true over a large part of the world that, in a given region and for a period of one or several decades, there has been one book regarded as the bird-watcher's "Bible." Such a book, in the days when I grew up, was Chapman's *Handbook*.[1] Anyone in eastern North America who took to the observation of birds had to arm himself beforehand with a pair of binoculars and this volume, which, in spite of its six hundred pages, could be slipped into a large pocket. While it could thus be carried into the field, its primary use was served by studying it before going into the field (as well as after returning). Not only did it provide—in its text, in its twenty-nine plates, and in its numerous text-figures—the information needed for identification, it provided much more as well. It told where each species bred and where it wintered, giving dates of arrival and departure at a series of points in eastern North America; it told where it was rare and where common; it described its nesting, and gave the dates of egg-laying at various localities. Above all, it presented a brief essay, perhaps half a page long, on the character, the "personality," and the behavior of each species. Here, for example, is its essay on the nighthawk, *Chordeiles minor:*

> In wooded regions the Nighthawk passes the day perched lengthwise on a limb, but on the plains he roosts upon the ground, where his colors harmonize with his surroundings. Soon after sunset he mounts high in the air to course for insects. Batlike he flies erratically about, and at more or less regular intervals utters a loud nasal *peent,* this call being followed by two or three unusually quick, flitting wing-beats.

---

1. Frank M. Chapman, *Handbook of Birds of Eastern North America* (New York, 1886, 1895, 1912, 1932).

Long after the light has faded from the western horizon we may hear this voice from the starlit heavens, for the Nighthawk is one of our few truly nocturnal birds. Occasionally the *peents* are given more rapidly, and, after calling several times in close succession, the bird on half-closed wings dives earthward with such speed that one fears for his safety; but when near the ground, or while still high above it, he checks his rapid descent by an abrupt turn, and on leisurely wing again mounts upward to repeat this game of sky-coasting. At the moment the turn is made one may hear a rushing, booming sound, which, as writers have remarked, can be imitated in tone by blowing across the bung-hole of an empty barrel. It is made by the passage of the air through the bird's primaries.

In late summer Nighthawks gather in large flocks and begin their southward migrations. When flying the white mark on their primaries is a conspicuous character, and has the appearance of being a hole in the bird's wing.

Picture, now, someone who has read Chapman's account of the nighthawk before ever seeing one. Because it has presented the bird so vividly to his mind's eye, when he first sees the living reality with his physical eyes he will enjoy the delight of recognition, and he will have an appreciation of it that he could not otherwise have had.

Chapman's *Handbook* did not confine itself to the appreciation of species only. Each family was introduced by a brief account of its basic character and its place in the scheme of things; as was each order. And the whole was preceded by an Introduction, 138 pages long, that constituted a short course in ornithology, under the headings, among others: The Study of Birds in Nature, Finding and Naming Birds, The Distribution of Birds, The Migration of Birds, The Voices of Birds, The

Nesting Season, The Plumage of Birds, General Activities of the Adult Bird.

✧

I, myself, came to know birds in the pages of books before I knew them in nature. As an undergraduate, I was captivated by the account William Beebe gave of the wandering albatross in his *Arcturus Adventure*. Immediately I had to know more about it, and then about the other birds that coursed the open seas on the far side of the world from where I should have been preparing for exams. I then studied W. B. Alexander's *Birds of the Ocean: A Handbook for Voyagers,* thereby acquiring a fair knowledge of the systematics of seabirds, their orders, families, genera, and species.

My classmate and friend, Charlton Ogburn (later a distinguished nature-writer), whom I consulted about these matters, was amused, but also a bit impatient, at the combination of my interest in absent birds with my indifference to those at home. He, himself, had been passionately interested in birds ever since, as a young boy, he had started going on bird-walks in and around New York City.

However, if my interest in birds began at the antipodes, it widened with time until at last it reached my doorstep. I acquired the two indispensable aids to birding, binoculars and Chapman's *Handbook*. During vacations in Westchester County, New York, where Charlton and I lived only a few miles apart, I began, in addition to birding by myself, to participate with him in the pursuit of birds. Out in the field during the daylight hours, day after day, I would in the evening study Chapman. We both came to know the *Handbook* so well that, at last, we could recite passages to each other from memory— sometimes antiphonally.

✧

Identification in those days was more difficult than now. The turning-point came with the appearance in 1934, two years after our graduation, of Roger Tory Peterson's *A Field Guide to the Birds,* the first of a series of field guides that have greatly facilitated identification, thereby attracting a wider public to the pursuit of birds.

The Peterson guides have, from the first, been confined to the sole purpose of identification. The idea behind them has been simple. It is that each form (adult male, adult female, or immature) of each species has some one or more marks, visible in the field, by which it may be distinguished from all other forms and species. In the pine siskin, for example, these marks are the yellow flash on the wing and the yellow at the base of the tail. In the Mississippi kite, as seen on the wing, it is the solid black tail.

Peterson, himself, never intended that his field guides should serve any purpose except that of identification. In his Preface to the first one ("Giving Field Marks of all Species Found East of the Rockies"), he wrote that it was designed simply "to complement the standard ornithological works." In fact, however, it became so popular that it displaced them. It became "the Bible" of all birders, in succession to Chapman's *Handbook,* which went out of print.

✧

Even the best revolutions involve loss as well as gain. So did the replacement of Chapman by Peterson, which was unintended by the latter. Identification gained, but the larger appreciation suffered.

Since the Peterson revolution, field guides on the Peterson model have come out for most of the regions of the world. On my own bookshelves I have such guides to the birds of Europe, the birds of South Africa, the birds of East Africa, and the birds of New Zealand. But there are others as well, and new ones keep coming out. They all have in common their confinement to identification.

Add to this that any number of other competing field guides, for virtually every region of the world, have now come out. Some of them have features that improve on the original Peterson, but they would never have existed except for Peterson's pioneering, and they share with his original the confinement to identification.

I must not give the impression that such information as Chapman provided has become unavailable since the *Handbook* was replaced by the field guide. On the contrary, the widening popularity of birding, to which the field guides have contributed so notably, has contributed in turn to the publication of a wide literature on birds. I think of the encyclopedic and spectacularly illustrated books, generally of large dimensions, such as *The World of Birds: a Comprehensive Guide to General Ornithology*, by James Fisher and Roger Tory Peterson (London, 1964); *Birds of The World: A Survey of the Twenty-seven Orders and One Hundred and Fifty-five Families*, by Oliver L. Austin, Jr., illustrated by Arthur Singer (New York, 1961); *A New Dictionary of Birds*, edited by A. Landsborough Thomson (London, 1964); and *A Dictionary of Birds*, edited by B. Campbell and E. Lack (Vermillion, S. Dak., 1985).

There are also the works devoted, respectively, to single taxonomic groups, of which the most impressive known to me is the two-volume, lavishly illustrated *Eagles, Hawks and Falcons of the World*, by Leslie Brown and Dean Amadon (New York, 1968). Excellent, although more modest, examples of this kind of thing are *Pigeons and Doves of the World* (London, 1967) and *Crows of the World* (Ithaca, N.Y., 1976), both by Derek Goodwin, and *The Swans,* by Peter Scott and the Wildfowl Trust (London, 1972). But books of this kind, invaluable as they are, are not designed for the mass of birders. None of them serve, in any large sense, the purpose served by Chapman's *Handbook.*

✦

In 1970 I spent eight days on an ice-breaker traveling from New Zealand to Antarctica.[2] Anyone who has observed birds at sea knows how frustrating it can be. Often one catches only brief and partial glimpses of an individual between the waves before it disappears forever. Or particular birds are just too far away, and, since the ship is not under one's command, one cannot bring her about to pursue them. Fortunately, I had a year before the trip to familiarize myself, in books, with all the seabirds of the Southern Hemisphere. (A year's preparation for eight days was about right in the circumstances.) During that year I not only learned what identifying marks to look for—in albatrosses, petrels, prions, and the rest—I also came to know, in preliminary fashion, the unique character of each species, together with its way of life and life-history. How hopeless it would have been for me to have consulted the books, for the first time, only at the moment of observation, or after the opportunity was over!

So I return to the point with which I began, that, for a full and true appreciation, one must go to the books before going to the birds themselves. This means that the proper kinds of books must be available, as Chapman's *Handbook* once was.

---

2. See my *The Sea and the Ice: A Naturalist in Antarctica* (Boston, 1973, and Ithaca, 1989).

A species is not an entity or distinct creation, but merely one link in the chain of bird-life, which, because of the loss of the adjoining link or links, appears to stand by itself.
— FRANK M. CHAPMAN[1]

It seems obvious to us that organic life is divided into species, such as the lion and the tiger, the mockingbird and the golden eagle, the Norway spruce and the white oak. But, when we come to define the term, we run into unexpected difficulties, and as we struggle with these difficulties we find increasing reason to doubt that, in fact, the concept of species is anything more than an arbitrary device for the classification of the variety with which nature confronts us.

According to the binomial system of classification, which is in universal use today, all life is divided into species that are grouped in genera. Thus the American robin, *Turdus migratorius,* belongs to a genus that has other species as well, such as *T. iliacus* (the song thrush) and *T. assimilis* (the white-throated robin). The convenience of this binomial system is clear. If someone tells me that a bird I see in East Africa is *Turdus pelios,* I immediately know of its kinship to other species with which I have long been acquainted. This is a basic element in my appreciation of it.

---

1. *Handbook of Birds of Eastern North America* (New York, 1932), p. 9.

A common assumption is that, even if genera, families, and orders are mere inventions of the human mind, species themselves are real. According to Genesis, species were created as such when "God made the beast of the earth after his kind, and cattle after their kind, and every thing that creepeth upon the earth after his kind." And when "Adam gave names to all cattle, and to the fowl of the air, and to every beast of the field," the names he gave were species names, for it was surely species that he was naming. The eighteenth-century botanist Linnæus, to whom the binomial system is attributed, living so long before Darwin, must have assumed that, whatever might be the case with genera, families, orders, etc., species had been created as such.

There is another established classification among us, that of subspecies (sometimes referred to as race, or simply as geographical variety). The members of a species will show minor variations, say in size or color, corresponding to the different geographical localities in which they are found, and this is the basis of subspecific distinctions, which are represented by an extension of the binomial system to make it trinomial. (E.g., the American robin is divided into *Turdus migratorius migratorius,* the subspecies that occurs in the northwestern part of its range, *T. m. achrustorus,* the one that occurs in the southeast, and *T. m. nigrideus,* the one that occurs in the northeast.) These geographical variations, however, are generally so slight that the tendency has been to accord less importance to the subspecific than to the other distinctions.

(Because all species tend to vary more-or-less continuously and almost imperceptibly over their geographical range, it is too easy to "discover" and name new subspecies. The excessive number of such "discoveries," which the taxonomists of the day would then name after one another, at last became a joke in the nineteen-thirties. Francis Lee Jaques, the great bird artist, propounded "Jaques's Rule," which is that the number of sub-species in any given area is inverse to the distance from the nearest natural-history museum.)

So it is that, although biolo-gists are disposed to recognize an arbitrary element in the con-cepts of order, family, genus, and subspecies, many have been reluc-tant to give up the concept of species as representing what na-ture created. The fact is, however, that one can formulate no definition of species to which exceptions don't have to be made, and this, in itself, is logically untenable. If I say that Longfellow and Emerson resembled each other in that both had beards, except Emerson, I have made what is a logically untenable statement. If I say that a species is defined by the fact that all its members, and only its members, can breed together to produce fertile offspring, and if I then go on to make excep-tions, represented by cases of interbreeding between the re-spective members of what we nevertheless consider to be sepa-rate species, I have invalidated my definition, for my exceptions contradict it.

The fact is that "species" has been most frequently defined as denoting a group of interbreeding organisms that are re-productively isolated from other organisms, with which they cannot breed to produce fertile offspring. But the exceptions are legion, some of them even representing interbreeding be-tween different genera. There is the example of a fertile hybrid produced by the mating of a common buzzard, *Buteo buteo,* and a northern goshawk, *Accipiter gentilis*,[2] and there are examples of hybridization between swans and geese.[3]

In *The Origin of Species,* and despite its title, Darwin wrote, "I look at the term species as one arbitrarily given, for the sake of convenience, to a set of individuals closely resembling each other."[4]

When we cannot satisfactorily apply our nominal categories to reality, the thing to do is to set them aside and proceed to examine the reality independently. Evolution is a continuous divergence of forms, sometimes proceeding slowly, sometimes fast, from a common ancestor. It is a process most conveniently represented by the diagrams that show what is called "the tree of life." Descendants of a common ancestor (shown as ascendants on the tree), becoming separated geographically, come to differ in consequence of their respective adaptations to different environments. Such differentiation or branching tends to be progressive with time, and we can, if we wish, regard it as proceeding from subspecies through the successive stages we call "species," "genus," "family," "order," etc. But the reality is of a continuous divergence rather than a succession of jumps. It follows that evolution is best represented by the diagram that, at its simplest, takes the form of the letter *V,* showing two branches diverging continuously from a common point of departure. (The tree of life is merely a spreading complex of such *V*'s.) If we accept the *V* as representing reality, we can see what is wrong with asking what degrees of divergence represent, successively, the series of categories that we call "subspecies," "species," "genus," "family," "order," etc.

---

2. Leslie Brown and Dean Amadon, *Eagles, Hawks, and Falcons of the World* (New York, 1968), p. 21.

3. Peter Scott and the Wildfowl Trust, *The Swans* (London, 1972), p. 86.

4. 6th ed., chap. 2 (New York & London, 1915), p. 66.

Let me offer an analogy. What our mercury thermometers reveal is an unbroken continuum from cold to hot. The degrees we mark along their lengths are no more than artificial devices of measurement; for the mercury does not, in fact, rise or fall by jumps from one degree to the next. So it is when we mark off the stages we call "subspecies," "species," "genus," etc. on the V of divergence. These stages, as such, do not represent a natural reality.

Isak Dinesen knew an old Danish clergyman who doubted that God had created the eighteenth century. He would have had equal reason to doubt that God had created either degrees centigrade or species.[5]

Nevertheless, the reluctance to see the traditional concept of species as an arbitrary convenience only, continues strong even in our day, a century after Darwin—if only because the division into species seems so real to anyone who goes out into the field and observes nature casually. (It seems so real because, as Dr. Chapman pointed out, the links that connect present-day forms with one another have been lost.)

If all the skins of all the birds that ever lived could be laid out along the lines of their evolution from the beginning to our own day, would we not have a perfect, unbroken continuum, from the archæopteryx to the song thrush, from the ostrich to the hummingbird, from every individual to every other individual? We would, it is true, see manifestations of abrupt mutations, such as might suddenly produce an albino. But such mutations would take time to spread through populations, if they did so at all, and would not, in themselves, constitute the advent of a new species. However rapid their spread, it would still require many generations.

---

5. *Out of Africa* (New York, 1938), p. 274.

One form taken in our day by the reluctance to downgrade the concept of species is an effort to show that the basic process of evolution is one of what is called "speciation"—which is to say, the generation of new species. Evolution, according to this thesis, does not proceed by an evenly continuous divergence, as represented by the letter V, but by something that would be better represented by a succession of steps, each marking the birth of a new species in the process of "speciation."

However, none of the "punctuationists" (as those who make this argument are called) go so far as to propose that a couple of one species may produce progeny belonging to an entirely new species, that the jump is from one generation to the next. When they suggest that evolution proceeds by quantum jumps that produce new species, they have in mind that, at certain times or places, under the forcing of rapid changes in the environment, new forms evolve so quickly that, in a rela-

20

tively short time, perhaps fifty thousand years or as many generations, they have become "new species." If this is the argument, however, then continuity of evolution remains, and so does the impossibility of providing an answer to the question of what point, in that continuous if rapid evolution, marks the emergence of a new species. (It makes no fundamental difference that the arms of the V are wavy.)

✧

A species may vary continuously in space, as in time, until at last it has to be regarded as constituting a different species. This was something Gilbert White could not have appreciated because of his lifelong abstinence from travel. The concept of species was more plausible to that contemporary of Linnæus because, in the seventy-three years of his life, he never traveled beyond a distance of perhaps thirty miles from his native village of Selborne, in the south of England. (What could he know of Hampshire, who only Hampshire knew?) Within his geographical and temporal confines, he would never have had occasion to observe a bird that could not be identified without question as belonging to one species or another. For example, along the beaches of Hampshire he would have seen, in each other's company, what were obviously two distinct species of gull, the herring gull, *Larus argentatus,* and the lesser black-backed gull, *L. fuscus.* The mantle or upper surface of the wings of the former were light gray, and its feet flesh-colored, while the mantle of the latter was almost black (quite black if he had observed it in Scandinavia) and its feet yellow. He would also have found that they did not normally interbreed. All taxonomists, even today, would agree that the two forms, as he saw them, were entirely separate species—"good" species, as they say. Now suppose, however, that he traveled widely. As he went eastward and southward, he would have found that the mantles of the herring gulls he saw became darker, and by the time he reached the Mediterranean he would have found that the "herring gulls" had yellow instead of pink feet. A point would have come (as it did for me on the island of Madeira) at which the gulls he saw were so perfectly intermediate between the two species he had known at home that he wouldn't have known to which they should be assigned.[6]

A plausible explanation of this continuity of change in space is that it is the visible representation of the continuity in time. We may suppose that an ancestral species occurred in the Bering Sea some twelve thousand years ago. Over the intervening millennia, its population spread eastward across North America and the Atlantic, and westward across Asia and Europe, until the eastward-spreading and westward-spreading branches met along the Atlantic shores of Europe and Africa. In the course of this expansion, the two branches, evolving differently in response to different environmental conditions, had at last come to differ so widely at their extremes as to constitute the two distinct species that we identify as the herring

---

6. David A. Bannerman, in his *Birds of the Atlantic Islands,* vol. 2 (Edinburgh), 1965, p. 43, referring to the form that breeds on Madeira, wrote: "Most recent ornithologists have never really agreed on whether it should be considered a subspecies of *Larus fuscus,* the lesser black-backed gull, or *Larus argentatus,* the herring gull. I . . . have now thrown in my lot with those taxonomists who treat it as a race of *Larus argentatus.*" Brought up as he was to the concept of categorical species, it never occurred to Dr. Bannerman that a form could be intermediate between two species, therefore not properly assignable to either one or the other.

gull and the lesser black-backed gull. In this case, however, the intermediate forms (the "adjoining links," to use Chapman's term) did not disappear, so that we are left today with a continuum between them, a gradation between two species in space, that visibly represents the continuum of their evolution in time.

The concept of species, as Chapman pointed out, depends on the disappearance of the intermediate links. When those links fail to disappear, the concept breaks down.

✧

Let there be no misunderstanding. I do not propose that we give up the concept of species, any more than I propose that we stop marking our thermometers with degrees. It is indispensable to the appreciation of birds—as are the concepts of genus, family, order, etc. But we should recognize the limitations that make its application to the living reality more-or-less arbitrary.

✧

There is reason to think that every organism on earth today, whether the reader of these lines or a microbe in his guts, is the product of continuous descent from a common ancestor of almost four billion years ago.[7] So all life-on-earth is one.

---

7. This evidence is briefly cited in my *Out of Chaos* (Boston, 1977), p. 160, footnote.

✦ III ✦

It is easy to recognize a swan. Swans are large, long-necked waterfowl with short legs and large feet, readily distinguished by the layman from all other birds. Unfortunately, if you ask any two or more taxonomists how many kinds of swan there are, the answers they give are unlikely to be the same.　　　　　　　　　　　　　　　　　　—HUGH BOYD[1]

However taxonomists may disagree on the swans of the world as a whole, all agree that Europe has three native species. One is the mute swan, *Cygnus olor*, with which we are all familiar because, in its domesticated state, it ornaments our parks. In Europe, where it is feral, it makes itself at home on the waters within the cities, expecting the tribute of bread from its human admirers. No other bird appears to have such a sense of its own value.

The other two European species, both of them wild and wary, are the whooper swan, *C. cygnus*, and Bewick's swan, *C. columbianus*. (They are considered conspecific, respectively, with the North American trumpeter and tundra swans.) Neither has the mute swan's habit of carrying its neck in a graceful curve, nor its habit of holding its wings out from its body as if it were an heraldic emblem. The two carry their heads high on straight stems and, at rest, keep their wings tight to their bodies.

---

1. In Peter Scott and the Wildfowl Trust, *The Swans* (London, 1972), p. 18.

Whooper swans that nest in Iceland fly five hundred miles over stormy seas to winter on the sheltered lochs of the Shetland Islands. (I mention Shetland in particular because it is only there that I have known them.) Bewick's swans that nest on the Arctic coast of Siberia fly across the Gulf of Finland and the Baltic, perhaps non-stop, to winter on the coastal fields of Holland. When such migrants arrive, being hungry and tired, they feed and sleep.

October 21, 1972, on the island of Texel off the Dutch coast, over twenty Bewick's swans, which had arrived in the preceding few days, were feeding or sleeping in a field. The spectacle was extraordinarily beautiful, because the sun was shining and the swans were in a flock of grazing widgeon that surrounded them on all sides and were interspersed among them. The contrast between the great size and whiteness of the swans, and the small ducks like wildflowers, gave a celestial splendor to the scene.

Suddenly there was heavenly music from the air and from the ground alike, and we saw four swans parachuting in to join the flock. They were singing as they sank through the air, and the swans on the ground, lifting their heads, were also singing to greet them. The four, feet extended, dropped lightly to the ground in the midst of the others, just as geese do when they land in a field.

After a moment's rest, the newcomers began to graze among their fellow swans and the widgeon.

✦

The wintering flocks of most species of geese graze in open fields, especially in the stubble of grain fields after harvest. To cope with the danger of mammalian predators, all such grazers have to be able to take off fast. The geese, heavy as they are, take off lightly and almost vertically, without a taxiing run; and it is common, on the Eastern Shore of Maryland (grazed by Canada geese) or on the fields of Holland (grazed by gray and barnacle geese), to see a dense flock suddenly separate itself from the ground as if by some gift of levitation, moving off like a cloud carried away by a breeze.

Whooper and Bewick's swans also graze in fields, although less commonly than geese; but this habit may be less natural to them. Seeing how lightly the four Bewick's swans drifted down to the field at Texel, I can believe them capable of an equally light take-off from land, but I have never seen one take off at all. Taking off from water is bound to require more effort, because the suction of the water has to be overcome. I have seen whoopers, which appear to be merely larger Bewick's (so much the two resemble each other), take off from water, but not without effort.

From its clumsiness in walking, by contrast with the other two, I judge the mute swan to be less adapted to land than either of them. Unlike them, it sheds all its grace when it moves from the water onto the land. Although it retains its air of dignity even under these circumstances, not aware that it is being ludicrous, it loses even that when it embarks on the long taxiing run, with frantically flapping feet and wings, by which it launches itself from the water. In the absence of a head-wind,

Bewick Swans

it needs the equivalent of a long runway for such take-offs, as also for the crash and skid of its landings. I cannot imagine one setting itself lightly down onto the land, like the four Bewick's swans. I once saw one take off from land, when threatened by a yapping dog, but this was on so steep a hillside that it hardly had to rise at all. I daresay that none of the three species were made for land, but the mute swan least of all.

✧

We expect birds to be nervous and flighty, easily alarmed and reacting instantly. The feral mute swans, however, seem to have no nerves and to know no fear. One of them, only inches away from me, where I stand on the shore, nevertheless plunges head and neck underwater to browse on the bottom. Or it curls its neck over its back and goes to sleep with me standing alongside. If I should shout, wave my arms, or threaten to seize it, it would simply move off as if injured in its dignity. Swans can have no fear of their inferiors, and that term covers all the rest of creation.

I daresay the mute swan, being so big, has had no occasion to develop a fear of predators under circumstances in which our own kind doesn't hunt it. It is too big to tempt an eagle or a fox, and it is no more exposed to wolves than to tigers.

Associated with its dignity is its indifference to the smaller waterfowl that crowd about it in such a city as Geneva: the coot, the mallard, the pochards and tufted ducks, above all the black-headed gulls. When the waterfowl are fed by a servitor of the Canton and Republic of Geneva at the Isle Jean-Jacques Rousseau, where the waters of Lac Leman empty into the Rhône, the coot, ducks, and gulls find themselves squeezed tightly in among the swans, body against body on all sides. The gulls actually land on the backs of the swans, and the coots sometimes scramble up on their backs. But the swans never notice. In the competition for the food that is thrown to them from above, they may peck at one another, but never at the lesser fowl, which peck at them and take food from their bills without attracting their notice.

Several times I have seen a gull riding around the harbor on the back of a swan, which appeared not to know it was there. (In Shetland, the black-headed gulls ride around on the backs of sheep.)

Mute swans react to us human beings because they come to us for their due of food, and of course they react to one another, but I can't recall one ever noticing any other kind of life.

✦

We may think in terms of two main groups of geese: those belonging to the genus *Anser* (together with the allied genus, *Chen*), and those belonging to the genus *Branta*.

Anser is composed of the gray geese, including the graylag from which our common domestic geese are descended. They are all basically alike, varying only in detail—some with yellow feet or bills, some with pink feet or bills, some having white on their faces, some being darker, some lighter, and one being blotched on the belly.

Branta is represented in North America by that remarkable species, the Canada goose, *B. canadensis*, with its black neck and head, and the white patch under its chin that extends to behind either eye. The chief representative of *Branta* in Europe is the barnacle goose, which grazes in its thousands on the fields of Holland, from which it rises as lightly as the gray geese—or as the wintering Canada geese on the Eastern Shore of Maryland.

At the Bosque del Apache Refuge in New Mexico, in late November, 1979, there were an estimated twenty-five thousand snow geese, *Chen cærulescens*, with a few Ross's geese, *C. rossii*, scattered among them; and thousands of Canada geese that included the largest subspecies, *moffitti* (which may weigh thirteen and a half pounds) and the smallest, *minima* (which weighs less than three pounds). Seen together, these two races appeared like St. Bernard dogs with Pekingese. I daresay the two races don't interbreed, but the survival of eight intermediate "links" makes their association in one species more plausible. (The largest of all, *maxima*, is now extinct.)

✧

The waterfowl constitute a more varied family than is suggested by the notion that they consist  of three distinct groups: ducks, geese, and swans. One might more plausibly  divide them into: the magpie goose and tree ducks; swans (of which one might plausibly be included with the geese instead); gray geese; black geese; shelducks or sheldgeese; dabbling ducks; torrent ducks; eiders; pochards; perching ducks and perching geese; scoters; harlequins, longtails, goldeneyes; mergansers and stiff-tails.[2]

One of the most familiar of these groups, which also appears to be one of the most coherent, is that of the dabbling ducks, so-called because they commonly get their food, not by diving in deep water, but by dabbling in shallow water. They tip up, tail in air, head and neck immersed and reaching down to the bottom, feet moving gently to keep them vertical. The mallard—best known to us in its domesticated versions, which are the familiar ducks of our barnyards and parks—is the typical dabbler.

---

2. Taken from Peter Scott, *A Key to the Wildfowl of the World* (Slimbridge, England), 1957.

Typical diving ducks like the pochard group (including the scaup, the American canvasback, and the European tufted duck), take off from the water by running over it, their wings whirring until they achieve the air-speed necessary to lift them above it—whereupon they retire their landing-gears under their tails. They can afford such a taxiing take-off because they feed on open bodies of water.

The dabblers, by contrast, take off in one vertical leap, from the top of which they proceed in normal flight. They have to be able to take off in this fashion because they so frequently feed in reed-beds or in narrow waters enclosed by vegetation.

Associated with these two modes of take-off are the relatively skimpy wings of the diving ducks—which are an advantage in fast and direct flight, during which they appear to oscillate like the appendages of some mechanism that has to be wound up—and the larger wings of the dabblers, which give them greater flexibility of flight if less speed. The dabblers are like the geese, which also have large wings to take off steeply.

The dabblers, unlike the diving ducks, are apt to come ashore and browse—as witness the widgeon that graze on the fields of Holland. Like the geese, they are good walkers, with strong feet placed under their center of gravity when they hold themselves horizontal. The diving ducks have their feet placed somewhat farther back, because the farther back the more useful they are for diving and underwater propulsion, although the less useful for walking. (Another family of water birds, the loons and grebes that can disappear instantly under water, have feet placed so far back that, on land, they have to hold themselves in a position approaching the verticality of penguins, and can't really walk at all. If they have to move on land, they can only shuffle on limbs that rest flat against the ground for their entire length.) When the diving ducks, whose feet represent a compromise, go below, they have to perform a little leap, as off a springboard, that involves arching through the air before disappearing into the depths. Grebes simply put their heads down and vanish.[3]

✧

Once, when rowing a skiff on the Lake of Geneva, I saw a tufted duck in the middle of a reed-bed, a position into which diving ducks don't usually get themselves. Wondering how it would take off without room to taxi, I approached until it did take off—straight up like a mallard.

Just as diving ducks can take off vertically if they have to, so dabbling ducks can dive if they have to. (Even swans can dive if they absolutely have to.) More than once, crossing the Lake of Geneva by motorboat, I have come upon an injured mallard that, unable to fly, dove to escape the menace I appeared to represent.

✧

The wild mallard is beautiful in its lines, trimmer and more neatly marked than its bloated descendant of the barnyard which, in the process of being bred to provide the maximum of flesh, has become like a fat man, as well as acquiring coarse and misshapen irregular markings. It is not altogether wrong to say that domestic and feral mallards, in their degree, are monstrous versions of the wild mallard, just as domestic geese, their bellies almost brushing the ground, are monstrous versions of the wild graylag.

---

3. Note that what we think of as the leg of a bird, the tarsus, which is generally vertical in the standing bird, is really part of the foot, the homologue of the joint from the heel to the toes in human beings. Birds walk on their toes.

Until a few years ago, the only mallards found on the Lake of Geneva, except occasional companies of farmyard birds on the shores of settlements, were the wild ones, with their elegant lines and delicate colors. The feral varieties, being sluggish and unwary, were shot by the hunters when they strayed into the open lake to join the wild ones, thereby being eliminated from the breeding stock. Then, however, all hunting of any sort was forbidden by law in the Canton of Geneva. Immediately, feral mallards began to multiply and spread unchecked, their progeny presumably breeding with the wild birds, to the point where today it is hard to find pure-bred wild mallards. The lake in the vicinity of Geneva, as in Geneva itself, is now crowded with the degenerate feral ones.

The cessation of hunting came as a relief to me personally, but it has had this one unwelcome consequence.

✦

Having in mind the fact that we oversimplify when we divide the family *Anatidæ* into three categories—swans, geese, and ducks—let us consider the smallest of the entire family, the so-called cotton teal, only 33 centimeters long, which makes it even shorter than that little ball of feathers, the pied-billed grebe of North America. It is a delight to see its companies on ponds in the forests of India, each individual a miniature of its general kind. But—are we right in calling it a teal, or even a duck? The shape of head and bill make it appear to be a goose, albeit one small enough for the mantelpiece. According to the taxonomists, it belongs to the tribe of "perching ducks and geese" (*Cairinini*), which includes one of the largest of all waterfowl, the spur-winged goose. And its alternative vernacular name, which seems to be preferred in the serious literature, is "pygmy goose."

✦

What generally presents itself to the mind's eye, when we think of the beauty of a particular species, is the individual bird. (So an admiring reference to the mute swan is apt to evoke the image of the single bird floating on the dark, reflecting surface of calm water.) In particular cases, however, we may be especially moved by the sight of large companies or swarms. I cherish, as one example, the memory of between a thousand and two thousand Canada geese grazing together on a field in Oregon, all facing one way and moving slowly forward, occasional sentinels among them, heads held high, alert on behalf of all. Or there is the recollected vision, already referred to, of the clouds of gray geese lifting off the fields of Holland as if lighter than air. Another example is that of redwinged blackbirds, their swarms against the sky looking like stretches of black chiffon coiling and uncoiling.

On a lesser scale, I think of the companies of black-headed gulls that flow over the housetops of Geneva, dipping and rising together, their wings moving to the rhythm of a dance. (By comparison, the spectacle of the ballet-dancers at the Grand Théatre of Geneva, trying to take off like birds, is awkward and pitiful.)

No bird is lovelier in its individual self than the little egret of the Old World (virtually indistinguishable from the snowy egret of the New), but the sight of its swarms passing in long and sinuous lines over the housetops of Katmandu has an appeal of its own.

Finally, there is the spectacle of such seabirds as the prions that, in their hundreds of millions, constitute a carpet of sparkling life from horizon to horizon of the Antarctic seas—and that of the great white gannets that fill the air like a blizzard off their nesting cliffs in Shetland.

✦ IV ✦

Since the beginning of civilization, there have been two worlds, the world of nature and the world of humankind. Until the most recent times, the latter was to the former as mere clearings in the wilderness. Wild nature was an encompassing menace against which a besieged humankind, in its little settlements, protected itself as best it could. Certainly there was no question of its protecting nature against humankind. We can be sure that our Crô-Magnon forebears did not worry about the possible extinction of the saber-toothed tiger.

In the past thirty thousand years, and especially in the past hundred, an accelerating expansion of civilization has, at last, reversed the situation. Today, in consequence, we are increasingly concerned to preserve the rapidly dwindling remnants of wilderness, and to save from an imminently threatened extinction its remaining large mammals and birds. The consequent rescue-operation has at last taken on the proportions of an emergency.

Although we must save as much of the natural environment as we practicably can, we are faced by the likelihood that, in default of either a miracle or a catastrophe, much of it will not be saved. We don't expect and we don't propose that the population of mankind be reduced to what it was even a century ago, let alone a thousand or ten thousand years ago. We don't propose to retract our civilization. We don't propose to dismantle our cities, our highways, or our farms. The most we can do is to limit future growth and future development. It is in these circumstances, then, that we are concerned to save so many species of the wild from a threatened extinction.

There are various ways of doing this. One is to set aside, where there is still time, such large reserves as the game reserves of Africa. This may be the only hope for some of the interdependent communities of big carnivores, ungulates, vultures, and eagles. In other cases it should be possible, without the establishment of reserves, to develop a new relationship between our own kind and the species of the disappearing wilderness, which we would save from disappearance as the rescuers in a rescue-ship would save the population of a ship that was sinking. The rescuers would set about finding accommodations for the refugees in their own ship. Here the problem is to find accommodations within our own civilization.

✦

It is certain that, in the time to come, some species will, without assistance from us, find ways of adapting themselves to the artificial environment of our civilization. The house sparrow—adapting itself first to our farms and at last to our cities—has done this  to such an extent that it would be threatened with extinction, today, if the wilderness returned to replace our civilization.[1]

The house sparrow is commensal with our kind; it feeds at our table. It has entered into this relationship of its own ac-

---

1. It is probable that it no longer exists anywhere except in close association with our kind. Where human settlements disappear, it disappears too. See D. Summers-Smith, *The House Sparrow* (London, 1963)—all of it, but especially p. 157.

cord, and we have neither welcomed nor objected to it. The point is that, in consequence, the sparrow is no longer threatened by the disappearance of wild nature, as its ancestors would have been.

In other cases we have welcomed and encouraged the commensal relationship, which could not have developed otherwise. I think of the eastern North American chipmunk, which comes into houses and sits under dining-room tables at mealtimes to catch the crumbs that fall or to receive hand-outs. (A colony, presumably descended from individuals that escaped from captivity, has become established in one of Geneva's parks.) Like the sparrow, the chipmunk, which can do so well in our urban parks and gardens, is not threatened by the disappearance of the wilderness. The same may be said of the gray squirrel in North America and the common squirrel in Europe.

✧

Size counts. A rhinoceros could not make itself at home in our midst as the house mouse has, and this would be only less true of moose or red deer. The golden eagle could not easily repeat the house sparrow's achievement in adaptation. But large mammals and birds can make the adaptation where they have the goodwill of our kind, albeit in some cases only with our deliberate help.

The mute swan is the outstanding example. Where it nests far from civilization, it is still a fully wild bird. For at least seven centuries, however, beginning in England, our kind has cherished and even pampered it for its beauty, to the consequent enhancement of our environment. On the Lake of Geneva today, it lives only in association with our kind, so that, along the remote stretches of shoreline where our kind is absent, it is absent too. The disappearance of our civilization, which might be the salvation of the trumpeter swan, would be only less of a catastrophe for the mute swan than for the house sparrow.

✦

One manifestation of our expanding civilization is the increasing urbanization of the earth's surface. The characteristic of our cities, by contrast with the towns that preceded them, is that they are no longer points in a rural landscape but spreading blotches that tend to merge with one another. London, for example, now occupies an appreciable portion of southeast England, being some thirty miles across, and beyond it lie extensive suburban areas. The rural areas, themselves, are constantly becoming less rural. Consequently, one solution to the problem of accommodating formerly wild species within our civilization is their urbanization, their adaptation to the urban environment—as we have been adapting ourselves to it.

I am constantly impressed by the degree to which this is being accomplished, even without human intervention, in the case of water birds and of middle-sized mammals. For example, even the center of London now has a permanent population of foxes, members of which have been seen in Trafalgar Square and elsewhere in its vicinity. Beech martins have made themselves at home in the warehouses of Geneva, where they live on rodents.

Perhaps the outstanding example is that of the gulls, which have proliferated as a result of their commensal association with our civilization—and these are, after all, birds much bigger than the house sparrow. Associated as they are with our civilization, they have experienced a population explosion like our own. Surely gulls were relatively rare in the days of the cavemen.

Except during their breeding season in spring, when they

resort to inland marshes, the black-headed gulls are primarily city birds throughout Europe. As I write, by a window that looks out over the center of Geneva, I see them everywhere on chimneys and the ridges of roofs. (If I should open the window, they would immediately fly to it in the hope of being fed by me.) The air is alive with their flying companies, large and small. They feed like pigeons amid the traffic of the streets, and they crowd the waterways, the harbor, and the Rhône River.

I say that the small black-headed gulls leave the city for nesting, but a number of the larger species of gull no longer do. In Britain, herring gulls have taken to nesting on the flat roofs of warehouses and factories; and those graceful gulls of the open ocean far from land, the seagoing kittiwakes, have taken to nesting on their window-ledges.

In the middle of the city of Narvik, Norway, in one of its central squares, with continuous movement of heavy motor-traffic and pedestrians around it, stands a small rocky outcrop preserved for the sake of the kittiwakes that nest colonially

upon its ledges. It is as if there were a kittiwake colony in the middle of New York's Times Square.

✧

The black-headed gulls that are so common on the waterways of Europe, although diurnal, nevertheless show a limited nocturnal disposition. Along the Rhône River, as it passes through Geneva, flying individuals or small companies are to be seen in the full darkness of night, illuminated from below by the street lights as they pass overhead. And the flood-lights trained on the cathedral towers occasionally show flocks of them passing over in the middle of the night—at least during the seasons of migration.

✧

In some cases, the human population of our cities seems not to be aware of what it harbors in its midst—or, at least, to know it for what it is. I think most people in Geneva, if they notice them at all, assume that the wild waterfowl in the city are domestic fowl of some kind.

Pochard and tufted ducks, which breed in the north of Europe, fly south to spend the winters in large numbers on the Lake of Geneva. Out in the open lake, where they have been hunted with guns for centuries, they are unapproachable. Every winter there is a flock of some two or three thousand in the middle of the lake, four or five miles from the city. For years I used to approach it by boat, as close as I could, in order to search its ranks, with my binoculars, for the occasional individual of a rare species in its midst; but I could never get closer than a third of a mile before the whole flock took off, like a cloud sliding along the surface—to land again a mile or two away. Inside the city, however, where they have learned that they are safe, these same ducks come to one's feet for food.

The principal bridge of Geneva, at the bottom of the little harbor where the lake empties into the Rhône, is the Pont du Mont Blanc. It trembles all day long under six lanes of roaring traffic, and its sidewalks are crowded with pedestrians. But all day long hundreds of pochards, generally asleep with bills tucked into their back feathers, are assembled on the lakeside of the bridge, many so close under its edge that one could drop pebbles on them.

I mention the pochard especially, but there are many waterfowl of other species among them. There are tufted duck and goldeneye. There are the innumerable coot (belonging to the family of rails and gallinules, rather than to that of the ducks), which are as domestic within the city as barnyard fowl. There are the little grebes, smallest of all, belonging to the family of loons and grebes—as also the great crested grebes, which one can watch swimming underwater when they dive to find food on the bottom. These and other waterbirds are so conspicuous that the pedestrians who cross the bridge can hardly fail to notice them, and, in fact, may stop to watch them. It is clear from their comments, however, that most don't understand what they are looking at. They are apt to think that the little grebes are baby ducks. A girl points to the coot and says to her boy-friend, "Look at those funny black ducks." Most of the passers-by would be astonished to learn that these were not domestic birds. The reality is greater than anything that occurs to their minds.

✦

The most beautiful of the waterfowl in Geneva (leaving aside the swans) are the goosanders—the same elegant species as, in North America, is called the common merganser. It belongs to a group of fish-eating "ducks" with slender, saw-toothed bills (to hold a slippery catch). It, and the other spe-

cies of merganser, generally hunt in fast-running water, where they dive repeatedly to catch their prey, but they are not uncommon on the still water of lakes.

Perhaps I should not have made an exception of the swans when I said that the goosanders are the most beautiful of the waterfowl in Geneva. The female, long and streamlined, is gray and white on the body, with chestnut neck, chestnut head, and a chestnut crest that always stands straight out behind as if she were facing a high wind. The even longer male is white with a black back and head, the white washed with pink, the black reflecting green.

The goosander nests in Iceland, Scotland, along the Baltic shores, and in Scandinavia. But there is also an isolated nesting population on the Swiss lakes. Since it nests in big hollow trees, its numbers have suffered from deforestation around the Lake of Geneva. Every spring one sees the females flying in and out among the trees of private estates along shore, looking for nesting sites but, for the most part, unable to find any.

Some years ago, however, a country doctor, Dr. Reynold Rychner of Versoix, had the inspiration to erect nest-boxes made of hollow logs at many places along the lakeshore. Almost all were immediately occupied (within twenty-four hours in the case of the one he erected where we were living), with the result that, over the next few years, the goosanders became fruitful and multiplied in our midst. (Was this not an excellent manifestation of what I am here advocating, the replacement of vanishing features of the wild by artificial equivalents?)

Then, a few years ago, an extraordinary accident occurred. A brood of six newly hatched goosander ducklings, mere brown-and-yellow powder-puffs, lost their parents or became separated from them. They immediately were "imprinted" on a pair of swans, which is to say that they took the swans to be their parents, following them wherever they went. (It was pure delight to see the two great swans, which paid not the least attention to their six tiny followers, swimming one after the other about the docks of Geneva, a file of goosanderlings ever behind them.) Adopting the attitude of their foster parents, the goosanders grew up with none of the fear of our humankind that their proper parents would have taught them by the example they gave.

In time, the six, grown up, had progeny of their own, to which they, in turn, communicated their trust in humankind. The result was the proliferation of a goosander population the members of which could be approached, perhaps by boat on the lake, to the point where one could have reached out to touch them—or nearly so. It follows that the goosanders, like the swans, are ready to accept the hospitality of our cities, if only we will extend it to them.

In fact, it has been extended to them in Scandinavia, where they are cherished for their eggs. Nest-boxes are set up for them along the shore by persons who then take from those boxes the first clutches of eggs, allowing the replacement clutches to hatch out. The result is goosanders that are tame because cherished. They swim about the busy docks of Oslo as fearlessly as if they were fully domesticated birds.

✧

I have mentioned that, every spring, one sees the female goosanders flying in and out among stands of big trees along the lakeshore, looking for hollow trees in which to nest. In the spring of 1980, one repeatedly explored the roof-tops and chimney-pots of the old city of Geneva (in the center of the modern city), landing on chimneys, poking her head into recesses that might serve for nesting.

I see no reason why such nesting, in the middle of the city, might not prove successful and become established—except for one problem. Under natural conditions, when all the young have hatched as balls of downy fluff, they tumble out of their arboreal nest in the woods, then follow their mother overland as she leads them to the water, which may be some distance away. Any ducklings that dropped from a roof-top nest in  the old city of Geneva would have to go a considerable distance through what are usually traffic-laden streets to reach the Rhône or the lake. However, I believe that the departure from the nest usually takes place early in the morning. At four-thirty in the morning, while the city still slept, this could be accomplished.

In fact, goosanders have now established themselves in the middle of the city of Neuchâtel, in nest-boxes that have been put up for them. Let the government of Geneva, and of all cities within the breeding range of the goosander, take note.

✧

Going down the Rhône from the lake, in the city of Geneva, there are six successive bridges. Between the fifth and sixth (the Pont de la Coulouvrenière and the Pont de Sous-Terre), the river narrows greatly, so that the water flows in rapids under a broken and undulating surface. On one side is the Quai du Seujet, lined with buildings, on the other—or, rather, on what is an island in the middle—is a great building full of the humming turbines of a power station. This rushing water, which harbors crowds of small fish, is especially attractive to the goosanders, who have a taste for turbulence. Consequently, in winter they crowd it so densely that I have sometimes fancied I could walk across it dryshod on their backs.

Is there any waterfowl more beautiful in its way than the drake goosander? The mute swan represents an overdone elegance, like something to be kept under glass, but the beauty of the goosander is the beauty of the wild.

If we deliberately bring in sculpture to adorn our city, why not also birds? Should not the municipal art-commission sit on the goosander?

I used to know the goosander in America (where it is called the common merganser), catching glimpses of it on woodland streams when it did not see or hear me first. I would never have thought, in those days, that it might be a city bird, laying its beauty at one's feet. The least it could expect in return is that we humans should appreciate it. But I see no sign that the pedestrians hurrying along the Quai du Seujet do. They have other things on their minds, such as the high rate of interest.

No doubt the commensal ducks and swans of the city, having become part of our civilization, now depend on the rate of interest as we humans do. If that is in fact the case, I would have the rate controlled and manipulated with not only ourselves in mind.

Perhaps a day will come when we think of a city as the abode of birds and people.

✦

Because goosanders have long, narrow bodies and long, narrow wings, when their squadrons fly over the river in formation, as if they were the Swiss air-force, they look like so many winged arrows. (The small, fast planes of the Swiss air-force sometimes remind me of goosanders.)

✧

Some city people don't go in for bird-watching because they think it requires trips into the country. Some who do go to the country weekends to look for birds, carrying binoculars and a field-guide, overlook the weekday birds in the city.

Taking a walk about the city before breakfast every morning, I see more birds, and more species of birds, than I would see on a like walk in the country.

✧

Few birds are more associated in our minds with the wilderness than the dippers, aberrant passerines the size of a thrush that haunt mountain streams and, diving down through them, walk at the bottom of the water along their floors. The species that occurs in the California mountains, *Cinclus mexicanus*, is the subject of what might be considered the most beautiful account of a bird ever written.[2] Its Alpine counterpart, *C. cinclus*, is just like it in the wildness of its habits and its habitat, and in its extraordinary behavior. It is the hardest of birds to observe—along its wild Alpine streams as in the Yosemite Valley—for it makes off on whirring wings even at the distant approach of a human being. But the first I ever saw was in the very heart of Geneva, on one of the stone supports of the Pont de l'Isle.

The Pont de l'Isle was destroyed by Julius Cæsar in 58 B.C., to prevent the Helvetians from crossing the Rhône, but has since been rebuilt.[3] What birds were found there in Cæsar's

2. John Muir, *The Mountains of California*, chap. 13, "The Water-Ouzel" (Boston, 1916). In the entire field of nature-writing, I know of nothing as good as this book, except for some of W. H. Hudson's essays.
3. See Julius Cæsar, *The Gallic War*, bk. 1, par. 7.

day? Surely few gulls, if any, but all sorts of marsh birds, with hawks and eagles to prey on them—including, surely, the white-tailed eagle, which we should invite to return.

✦

Because birds that are fearful of our kind in the country may have no fear of it in the city, should we not take advantage of this, if not to send out invitations, at least to welcome and facilitate their visits? Perhaps the tourist bureau should include them among the groups to be attracted.

I once saw a golden eagle pass high over Geneva on motionless wings.

✦

We must hope to limit the macadamization of the world, but, to the extent that we do not, we must hope that birds and mammals of the wild can become adapted to living in such a world, as they had to become adapted to living in the world of the ice-ages.

✦

Implicit in these pages, as in our customary thought, has been a distinction between wild and domestic, whereby all our Western birds and beasts are identified as belonging to the one or the other. On the Indian subcontinent, however, where the traditional Hindu respect for animals and birds has brought much of what would otherwise be wildlife to the condition of the mute swan in the West, the distinction becomes uncertain. So it is that the "wild" monkeys, langur or rhesus, are found in the hearts of cities, begging alongside the human beggars, not hesitating to pluck at the clothes of those they solicit—or even to jump upon their shoulders if allowed to do so.

Might one not say, under these circumstances, that such species can be divided into wild populations on the one hand, and commensal on the other?

It is generally impossible, however, to draw a clear line between the one and the other.

Consider the peafowl, the cock more ornamented than royalty itself. One may come upon individuals or small companies in wild forests where their behavior is wary, like that of any wild species. But one also finds them in the crowded centers of towns, where they feed at one's feet. In these circumstances we might suppose that there is, on the one hand, a wild population and, on the other, one that is domestic or commensal. In fact, however, no line can be drawn between the one and the other, for the farmlands and fields that separate the towns from the wild forests are continuously populated by their kind in a state of what we may call intermediate wildness.

One might make a comparison, here, with the foxes of England, some of which have become completely urban, inhabiting the center of great cities. But the foxes live in the cities clandestinely, retaining the wildness of their ways, whereas the town peafowl of India manifest no more wildness than barnyard fowl.

Like the peafowl, one may see that daintiest of birds, the little egret, feeding in the streets of a crowded village where it is all but jostled by inattentive human beings, as well as by dogs, pigs, and cattle.

I have referred to the house sparrow in the West as the prime example of a species commensal with our kind. But in India it enters restaurants and the lobbies of hotels to forage. (Indeed, a population that appears to be permanently cut off from the outdoor world inhabits the main building of Delhi's airport, where it no longer experiences the distinction between night and day in terms of alternating darkness and light.) The common mynahs and the jungle babblers—those lovely thrushlike birds with their white eyes and long, loose tails—also enter restaurants to forage over and under the tables.

✧

Certain large birds have constituted themselves a Department of Public Sanitation in the cities of the Indian subcontinent. The pariah kites and the vultures are relatively scarce in the open country, but in the sky over any human settlement one may see half a hundred vultures and an equal number of kites, all moving together in interlocking circles. Vultures and kites alike roost overnight in the trees that line the streets of towns and cities, indifferent to the traffic below them. Normally the vultures don't take wing in the morning until the sun, by heating the land, has engendered the rising masses of air on which they soar.

In the center of New Delhi one may, when one rises in the morning, look out of one's hotel window at the vultures in the adjacent trees—white-backed vultures, king vultures, and long-billed vultures together, perhaps still asleep, their heads under their wings. When one sees them with heads upraised, they appear to be survivors of antediluvian times—as if pterodactyls, still inhabiting the earth, had taken to our cities. Hunched up as they usually are, their naked heads supported on naked stems, they might be the spirits of old men. Their large eyes make them seem indifferent, disenchanted, and baleful.

✧

The garbage heaped up in the streets of the Indian towns and villages invites a wide variety of animal and bird life, competing for what it offers. Geneva, by contrast, must be one of the cleanest cities in the world, its streets scrubbed every morning before the citizen's day begins. Might not the city government consider, however, the establishment of garbage-heaps at intervals about the outskirts to attract the great vultures that once were common in the region, and that could be brought back?

I recognize that my suggestion, being impractical, is not to be taken seriously. To begin with, one could hardly get a good press for vultures— even though, in flight, circling and tilting overhead, they are so beautifully buoyant—each like a single broad wing with upcurved pennæ at either end.

In fact, the last of the great vultures of Europe, the griffon vulture and the bearded vulture (also known as "lammergeier," although it does not prey on lambs), disappeared from the Alps adjacent to Geneva only at the end of the last century, and efforts are now being made to reintroduce them. The small Egyptian vulture, beautiful on the wing with its black-and-white plumage, and its bright yellow head, still occurs in the south of France (as outside hotel windows in India) and has been seen in the vicinity of Geneva. (Might it not have been a garbage-collector in Geneva before the days of modern sanitation?)

✦

According to Brown and Amadon, the pariah kite "is one of the world's most successful and numerous birds of prey, probably the most numerous raptor of its size anywhere, and in some areas incredibly common."[4] Its cosmopolitanism, however, is disguised by the fact that it has been given different

---

4. Leslie Brown and Dean Amadon, *Eagles, Hawks, and Falcons of the World* (New York, 1968), p. 265.

vernacular names in the different regions that its respective races occupy. Thus in Europe and Africa it is called "the black kite" (although it is not black), and in Nepal, where it is lighter than elsewhere, it is called "the dark kite." But it is all one species, embracing alike the birds that swarm over the Indian settlements and those that, arriving from Africa in March, swarm over the rooftops of Geneva and over its adjacent lake.

Brown and Amadon, remarking that in some places this species "occurs in thousands, breeding in trees in the streets, and obtaining food from the roadways and back-yards," note that "it will snatch food from tables of market stalls in the immediate vicinity of the protesting owner." In the middle of Delhi, I myself saw one, in a long and lovely stoop and swoop from on high, snatch an offering from the uplifted hand of a little girl.

Those familiar with Kipling's tales of India will recall Shir the Kite, in *The Jungle Books*, and the kites in "The City of Dreadful Night."

✦

Until sometime in the nineteenth century, kites, along with ravens, were the common scavengers of London—well known as such to Shakespeare, who refers to them repeatedly. While one must respect the conclusion of the ornithologists who now identify their species as the red kite, *Milvus milvus*, of which a remnant population exists in Wales, the smaller and more agile black or pariah kite, *M. migrans*, still numerous in summer just across the Channel, seems a more likely bird for that role—except that it is absent in winter. The two species are so similar that, conceivably, the English of former times, lacking field-guides and the habit of making distinctions, did not distinguish sharply between them.

✦

The masses of people in India, crowded together under circumstances in which their lives do not rise high above the elemental level, are not as far removed from nature and the other animals as the people of New York City or Geneva. Moreover, they may not be as unhappy as, considering their poverty, we expect them to be—just as the medieval European peasants in the paintings of Brueghel the Elder appear happier than, considering their condition, we would expect them to be. In any case, one thing from which neither the Indian masses nor Brueghel's peasants appear to suffer is loneliness. Their problems are unlikely to be those of lonesome individuals, as is so generally the case in the developed societies of the West, but must be in some degree the common problems of the communities in which they live. Like the mammals that live always in herds and the birds that live always in flocks, they are never alone during the normal course of their lives.

My particular point here is that, even as a species, they are not alone. A visitor from Mars would not see settlements of humans but settlements that embraced humans, cattle, buffaloes, pigs, dogs, and a wide variety of birds, all living pell-mell together. (He might, indeed, be impressed by the fact that the zebu cattle, wandering through the crowded streets, generally require the humans to get out of their way, using their horns if necessary—thereby showing themselves to be at the top of the peck-order!) In any case, the mutual tolerance between people, cattle, dogs, pigs, monkeys, kites, vultures, egrets, mynahs, crows, etc. is notable to the visitor from the West. (One thinks of Joseph Wood Krutch whispering to those little frogs, the spring peepers, "Don't forget, we are all in this together.")[5]

---

5. In *The Twelve Seasons* (New York, 1949), p. 13.

The crowding and mixing of species bear on the obscurity of the distinction between wild and domestic.

✧

It is curious, but not irrelevant, that, although the elephant is the principal beast-of-burden employed by the human population of the Indian subcontinent, no domestic variety exists, the animals to be domesticated being recruited by capture from the wild population. (Where else is anything like this true?) *falconry*

✧

The small house crows of the subcontinent—half black, half blue-gray—appear to be as commensal as the house sparrows in their relation to our kind, even entering houses. But they have close relations with other species as well. For example, they ride around the towns on the backs of cattle, pigs, or goats. (They ride around the countryside on wild boars, deer, or antelopes.) Where a king vulture and a long-billed vulture were feeding on some carrion, individuals from a rabble of these crows, standing by, repeatedly jumped up onto their backs to poke their beaks under their back feathers—searching for what? When the vultures drove them off by threats with their open bills, they promptly returned to resume whatever their search was.

✧

In the wild forests of the Indian subcontinent one may hear the most familiar of domestic bird-sounds, the crowing of a rooster. The red junglefowl is the wild species from which our common barnyard poultry are descended. It is a woodland bird that does not readily come out into the open. Unlike their gallinaceous relatives, the peafowl, these bantam fowl appear not to have taken of their own accord to human settlements— but I have seen them, individual cocks or hens, approaching with the utmost alertness and circumspection human settlements in the forest where food was put out for them. The wild rooster bears feathers that are ornamented primarily in red, orange, and yellow, its laterally compressed black tail crowned with the sickle-shaped central feathers that we see in the breeds of its domestic kin.

Mankind's selective breeding of *Gallus gallus* for the ostentation of its beauty has accomplished wonders, as Darwin noted, but without improving on the original as one sees it in its native setting.

✧

The common mynahs, those starlings of the East, walk with long strides of their yellow legs about one's own slower feet. Little doves, ringed or spotted, with short feet but quick of wing, pass in small flocks like grapeshot. The white-breasted kingfisher perches on a wire above the various species of villagers, attentive to catch insects in the village ditches or, like a flycatcher, in the village air.

It is clear that, in southern Asia, our kind has not detached itself as far from other creatures as in the cities of the West. (Yet the cities of the West, too, are alive with birds.)

✧

The city of Jodhpur is overlooked by a high wall of crags, which is extended upward by the battlemented walls of an immense fortress. On top of the fortress, in turn, is a heavenly kingdom in the form of a complex of marble palaces, their bays and balconies carved like jewelry. These bays and balconies, looking outward in all directions, protect the privacy of inner courtyards at different levels, the whole suspended above the mundane city so far below. The Hanging Gardens of Babylon, themselves, could hardly have been closer to Heaven. This would be a fit setting for the great-winged angels that are said to surround God's throne.

On one of the palace ledges, looking out over the endless

plain with its little human habitations, one of the great-winged vultures sits upon a nest, covering eggs or young, only occasionally lifting her head from under her wing to check upon the surrounding world. No doubt her family has occupied these heights since long before the days when they were co-occupied, however briefly, by Akbar the Great.

Two kites, occasionally pausing to scream at us invading humans or to drive off other kites, are carrying sticks to a nest they are constructing on a ledge over one of the high court-yards. (Presumably their ancestors, lacking a proper respect, screamed as freely at Akbar and his courtiers.)

It is clear that the population centers of India do not belong exclusively to our own species.

✧ V ✧

The most ærial of birds, those that come closest to living their entire lives on the wing, are the swifts, of which there are some sixty-five species the world around. They spend their days coursing the skies, sometimes high, sometimes low, their mouths open to take in the swarms of insects that are their entire nourishment. Like swallows, they drink while skimming over the surface of still water. Unlike any other birds at all, they copulate on the wing. Some species build nests entirely of their own saliva, which hardens like glue, so that they don't have to gather nesting material. Others pick up detritus in flight, which they cement with their saliva to make egg-cups fastened to walls.

Out of the air, swifts are but crippled creatures, their feet incapable of anything more than clinging to a vertical surface, their wings too long for folding into their bodies. A swift accidentally grounded may find itself unable to take off and regain the air. As the fish's medium is the water, so the swift's medium is the air. Grounded, the one is as pitiful as the other.

Except for a few swallow-tailed species, swifts appear neckless and tailless in flight, like bullets with hinged wings.

In virtually all birds, the visible part of the outstretched wing has one joint, which divides it into an inner and an outer segment. In swifts, however, this joint is so close to the body, where the wing sprouts, as to be hardly noticeable. All one sees in the flying bird is the outer segment, which is long, narrow, and curved like a scimitar. So, being jointed only close to the body, the wings oscillate inflexibly, or are held rigid for gliding. The swift on the wing seems more mechanical than organic. One could imagine winding it up and launching it from the hand.

One does not normally see a swift alone in the sky. They swarm like insects, so that one can never count them. They sweep in circles, milling about on interlaced trajectories so that one can hardly follow the course of a single bird.

Birds so specialized for life on the wing could not enter into intimate relations with such crawling creatures as ourselves. One could not keep a swift as a pet. Although one holds the live bird in one's hand, the blankness of its eyes seems to preclude such communication as one might have with a parrot, a duck, or a canary. The swift in the hand is no longer a swift. I suspect that swifts inhabit a world of birds and insects, in which people are marginal. (No swift, I am sure, has ever noticed me.) Nevertheless, having adapted themselves to our civilization, they are among the most abundant of our city birds.

✧

No one can know how common or rare the chimney swift, *Chætura pelagica,* was in North America before the white man arrived to transform the continent. The only accommodation it found for overnight roosting and for nesting was in large hollow trees, such as could be entered from above by the swift swarm. Its numbers would have been limited by the availability of such trees. We may plausibly suppose that, with the deforestation that accompanied the spread of the white man's civilization, the number of swifts decreased—perhaps to the point where they became relatively rare.

However, there are pioneers among birds as among men.

Some individual bird, perhaps governed in its behavior by a misplaced gene, will adopt a way of life previously unknown to its kind; and its progeny may follow suit. I picture a particular swift that, on a particular occasion, descended into the chimney of a house as if it had been a hollow tree. Others followed its example, hollow trees having become so scarce, and, once they began to nest in them, chimneys would have seemed the only proper nesting sites to their progeny when they, themselves, were ready to nest. Within a few years, chimneys would have become the only nesting sites of these swifts in North America—as is the case today. The result would have been a reversal of the decline in their numbers—until, like the gulls, they had become more abundant than they had ever been in the aboriginal wilderness.

All summer long now, the skies above American cities are swarming with swifts, which, no less than the house sparrows, are city birds. As the gulls are scavengers of the city's waterways, so they are scavengers of the city's skies, which they sweep clean of insects.

✧

The chimney swift belongs to the genus *Chætura*, the spine-tailed swifts, so called because the shafts of their tail-feathers are adapted to propping them, like woodpeckers, against the walls to which they cling. Another widespread genus is *Apus*, some or all of its species having tails that, however short, are forked.

The European counterpart of the American chimney swift is *Apus apus*, the bird that the British call, simply, "the swift."[1] The common swift, as I shall call it, is a more impres-

---

1. The British (God bless them!) have traditionally thought of the birds they know in their islands as being British. Therefore, in naming them, they

sive bird than the chimney swift—larger, swifter, less fluttery in its flight, and appearing black rather than dark brown. There is evidence, almost conclusive, that its swarms may spend all night as well as all day on the wing. Originally it nested in holes or cracks in the walls of caves or cliffs; but it has found that city buildings supply all its needs in this respect, and so it, too, has become a city bird.

✦

Speed depends on momentum, which depends on weight, and weight tends to go with size. No kind of propulsion would enable a mosquito to keep up with a swift. So big birds, as a rule, fly faster than small ones. A golden eagle gliding tranquilly across the sky, even though it doesn't seem to go fast, goes faster than a house sparrow exerting itself. (If it doesn't go faster than a swift, that is because it is not specialized for speed as the swift is.) So the swiftest of the swifts are also the biggest. They are the most impressive for size and momentum alike.

The little chimney swift, the common swift, and the big Alpine swift have the same basic shape, that of the bullet with sickle wings. But the Alpine swift is more than twice as heavy as the first, more than a third heavier than the second. It is one of the fastest birds in the world, perhaps the fastest. It is a shark of the sky, with its white throat and belly, with its long curving fins that whip the air. Paul Géroudet describes with

---

have not felt the need to add adjectives by which to distinguish them from "foreign" birds. Since there is only one swift in Britain, it needs no adjective. This obsolete custom is beginning to break down as British birders extend their birding across the Channel, but with remarkable slowness.

what ease its swarming numbers pass back and forth overhead—"but, when they descend to the level of the flowered
meadows, their lightning passage leaves the common swifts, in
their evolutions, far behind."[2] The flash-past of an Alpine
swift matches the intake of one's breath, lasting no longer.

Like the common swifts, this species has come to establish
its nesting colonies in the cities, building its nests of saliva and
detritus under eaves or in church towers. The big birds pass
over the rooftops of Zürich like shots from a sling, curving in
their trajectories, trilling in chorus as they go . . .

✧

The chimney swift is only one example of a bird's adaptation to human surroundings. The pileated woodpecker, a
strikingly marked bird the size of a crow, appears to have been
making such an adaptation within my own lifetime. Originally a bird of the great forests of North America, it had for
centuries retreated before the advance of the white man with
his axe. So it was that, in my own youth, it was a species,
largely confined to the remnants of the forested wilderness,
that one could hardly expect ever to find in suburban surroundings. After the first quarter of the present century, however, it evidently began, quite suddenly, to adapt to our human
world—until, in a short time, it was nesting even in the city of
Washington's Glover-Archbald Park.

One speculates that, at a certain point in time, a pair of
woodpeckers with less than the normal wildness or fear of human surroundings, nested and raised young in proximity to
human habitation and activity. The young they raised would
then have accepted the environment in which they had been

---

2. Translated from *La Vie des Oiseaux: Les Passereaux* (Neuchâtel,
1951), vol. 1, p. 42.

raised as natural, and would consequently have chosen it for nesting in their turn—thereby producing an ever expanding population that represented a successful adaptation not unlike that of the swifts to chimneys.

✧

Among the most widespread and conspicuous of European birds is the European blackbird, *Turdus merula*. Related as it is to the American robin, *T. migratorius*, it occupies a corresponding niche in European suburbia, hunting worms on European lawns as the robin hunts them on American lawns. It runs suddenly across the grass, stops suddenly, cocks an eye, and suddenly stabs the earth with its bill to catch a worm, which it then pulls out of the ground—just like the robin.

It is, however, more ubiquitous than the robin, nesting

abundantly in the dense center of cities, where it may build its nest under balconies, in recesses of stone or cement.

Add to this that it is, by common consent, the finest singer in Europe after the nightingale. On spring evenings, our European cities are enlivened by its continuously caroling song—which may, indeed, be heard in the fullness of night, under the lamplight of Geneva, all winter long. In Britain and on the Continent, it is, surely, among the most abundant and conspicuous of birds. How is it, then, that it does not figure in English literature of past centuries—in Chaucer or Spenser or Shakespeare, in Milton or Cowper or Wordsworth, or in any other poets before the late nineteenth century—poets who, for the most part, wrote so eloquently about the nightingale and the lark, about the throstle and other songbirds less common and less conspicuous than the blackbird is today?[3]

The fact is that, until sometime in the nineteenth century, the blackbird was generally confined, in its habitat, to the dark forest, to which its blackness, making it inconspicuous, was adapted. (In its present habitat, its blackness makes it particularly conspicuous.) Then, however, as I have supposed to be the case with the chimney swift and the pileated woodpecker, it became adapted, quite suddenly, to the artificial human surroundings and, in consequence, enormously increased its numbers.[4]

May this not, in the future, happen with other species as well?

---

3. The reference in Shakespeare's *Midsummer-Night's Dream* (3.1) to "the ousel cock, so black of hue, with orange-tawny bill" may be an exception.

4. See L. A. Batten, "Population Dynamics of Suburban Blackbirds," *Bird Study* 20, no. 4 (December 1973).

✧ VI ✧

If someone tells me that a bird is a species of "swallow," I know what he means. The term refers to all the species of the family *Hirundinidæ*, and to them only. But when I read, as I often do, that a bird is a species of "eagle," I confront the fact that I don't know what an eagle is. What does the term mean?

The diurnal birds of prey, grouped as the order *Falconiformes*, are divided into seven families, including the *Accipitridæ*, which has sixty-four genera totaling 217 species. One of the sixty-four genera, *Aquila*, includes in its nine species *A. chrysætos*, the golden eagle, which occurs in mountainous regions throughout the temperate Northern Hemisphere. In legendry, this eagle has been to the birds what the lion has been to the beasts. If anything is an "eagle," it is.

No one doubts that the model for the eagle on the ensign of the Roman legions was *Aquila chrysætos*, and that the same species was the model for the heraldic eagle on European coats-of-arms. *Aquila* is the Latin for eagle, and *chrysætos* (which means golden) is clearly the type species—what our forebears had in mind when they referred to "the eagle." One may go on to say that *Aquila*, with its nine species in all, is the typical eagle genus—the only genus of "true eagles," if one uses the term in the strictest sense. The trouble is that, in practice, the term is applied, as well, to almost a third of the 217 species in the family *Accipitridæ*, diverse as they are.

Best known among the other so-called eagles is the American or bald eagle, *Haliæetus leucocephalus,* also a heraldic character. Except that it is in the same large family, *Haliæetus* is unrelated to *Aquila,* belonging on an entirely separate limb of the family tree. The eight species of *Haliæetus*—all of which have, in the adult plumage, white heads, white tails, or both—are better called "fish eagles" (or, as is common, "sea eagles"), for all live largely on fish that they find, dead or alive, along seacoasts or water-courses.

How about the many other species that are officially called "eagles"?

There are, in the family *Accipitridæ,* some fourteen species, generally very large, that bear the name "vulture" and may be regarded as a distinct group. Aside from them, all the birds that are above a certain size are called "eagles," despite the absence of any corresponding relationship among them. The problem this creates in North America is minimal, since it has just the one eagle and the one fish eagle. It creates a greater problem on the other continents. In Europe, for example, one comes across birds, not particularly aquiline, that seem to be large versions of the buzzards (genus *Buteo*) but bear such names as "Bonelli's eagle," "booted eagle," and "short-toed eagle." The observer is bound to ask what makes them "eagles," rather than just "hawks," and the only answer is that they are called "eagles" merely because they are above a certain size. Any large diurnal raptor, if it is not a vulture, is called an "eagle."

I like to think that Adam did better than we have when, at God's bidding, he named the species.

✦

Of the four phenomena that Solomon found beyond his comprehension, one was "the way of an eagle in the air."

The eagle that the author of *The Proverbs* was most likely to

have known in nature was the golden eagle. It is a mountain bird that, where I live in the Swiss Alps, normally hunts above tree-line. Its habit is to follow the contours of the broken slopes, gliding swiftly over grass and rock on wings that have no need to beat. In this fashion (but I have never seen it actually happen myself), it surprises its prey, such as marmot or ptarmigan, emerging suddenly from behind the silhouette of crag or slope to pounce upon it.

The shark in the seas is not more buoyant. Once, for ten minutes, I watched an eagle passing back and forth along a gentle slope above me, through a thin stand of spruce at the limit of trees—ever back and forth, close to the ground, banking at the turns, without ever a wing-beat.

All the great inland soaring birds (we might tell Solomon) depend on a rising air-mass for their support: either the drift of air against a slope that deflects it upward, or the rise of air that has been warmed by sunlight on the ground below. There is an equation between the force of the rising air, on the one hand, and, on the other, the combination of the bird's momentum with the extent of surface it opposes to the air. The bird's momentum, in its turn, depends on its weight, as the surface it opposes to the air is related to its size. Such an equation, such a balance, could not be established in the case of a butterfly, for, lacking sufficient weight, it would be carried helplessly aloft. In the case of a sparrow, such an equation could not be established because, lacking a sufficiency of weight, together with a matching surface, it would fall to the ground in such air as supports an eagle. So it is that the possibility of soaring is limited to birds of great size.

Does not the beauty of the spectacle presented by the eagle reside in its illustration of natural law—the laws of mechanics, in this case? The equation one intuitively apprehends, as one observes the spectacle, has the balance of a work of art. Indeed, what one sees in the spectacle is what one hears in the music of Bach, its combinations of rhythm and pitch to produce an overall balance. (I, a writer rather than an eagle or a composer, would like to match by the balance of my phrases the way of the eagle in the air.)

The flight of the eagle, illustrating natural law, is beautiful as Einstein's theory of gravitation is beautiful. (Tell Solomon.)

✦

One summer day, I was standing just above an Alpine col when, looking down, I saw a golden eagle circling up from the vapors of the steep valley directly below me, as if rising out of Hades. In a minute, growing larger, it was just beside me (immense at that range), and in another minute high above, becoming quickly smaller until, having risen high enough, it was merely a flake drifting across the blue toward the snow-cap of an Alpine peak.

Another time, in winter, I stood in the snow at the top of a cliff while a golden eagle glided past immediately below me, so close that I could see the individual feathers of its back and every golden hackle.

In a high bowl of the mountains, a golden eagle was playing in the sky with the limp and tattered corpse of an Alpine hare. It would repeatedly let it fall from its talons, as if it had no further use for it—but suddenly, when it had fallen far enough, would turn toward it and, at the end of a long curving stoop, retrieve it, rising with it again on slowly beating wings. It circled the bowl repeatedly like an ice-skater circling a rink, performing all sorts of evolutions as it went, sometimes with the dead hare, sometimes without. Repeatedly the hare, released, fell like a rag doll, and repeatedly it was retrieved before it touched ground. In some of the tight turns the eagle executed, it banked to the vertical. Once or twice it banked beyond the vertical, so that it was over on its back.

Another time, I climbed above tree-line with an English visitor, who told me that one of his unfulfilled ambitions had been to see a golden eagle. A moment later, three emerged from behind a mountain that rose above us, flying in file. As soon as they were overhead, one of them put on a display of aerobatics, turning over on its back, tumbling wing over wing.

Is there anything shown on the stages of our city-theaters that equals what nature has to show—even if only on rare and unexpected occasions?

✧

The golden eagle so masters its element as to give the impression of superfluous power. Sails spread, it hangs in the air as a fish in water— or, without effort, moves across it from horizon to horizon at a speed a racing pigeon could hardly match.

A common buzzard ("buteo," in American usage) was soaring serenely over the forested slopes that border the upper Rhône Valley. Suddenly, as if struck, it turned head downward and, wings beating, plunged into the forest. A moment later an eagle emerged overhead, wings spread and motionless. It continued across the valley, with its villages and farms a mile below, not swerving from its course until, half a minute later, its dwindling form had disappeared against the mountains on the opposite side. The buzzard and I both recognized, each in his own way, that this was the king of the birds.

✧

The American bald eagle is as big and powerful as the golden eagle. Indeed, it is heavier, its equally great wings extending straight out like broad planks. But the golden eagle combines its power with a grace that must reside in the refinement of its entire silhouette, especially in the subtle curves and flexures of its wings. Carried just above the horizontal, the wings slope up at the ends, a design that facilitates tilting. I daresay that, moving so close to mountain-slopes as it does, it has to have this facility. The bald eagle, a bird of coasts and wide waters, has less occasion to move through canyons and chasms.

The turkey vulture has equal grace, tilting and rocking in its flight, but gives no such impression of swiftness and power, neither of which are necessary to its way of life.

✧

Our perceptions are inevitably influenced, if not determined, by certain value-judgments that we have traditionally made. Thus the lion, a hunter, is a noble beast, while the jackal, a scavenger, is ignoble. If the eagle, in English literature, has been the noblest of birds, the kite, also a scavenger, has been the symbol of ignobility.

The references to kites, especially in Shakespeare, assume in the reader the familiarity that breeds contempt. This seems odd in our day, when the only kite that occurs in the British Isles, the red kite, is reduced to a remnant population in a small corner of Wales. In past centuries, however, this same kite was the scavenger of English towns, a common street bird in London. It was the garbage-collector, along with the ravens that were also city birds in those days. I daresay that those who served behind stalls in the open food-markets of Shakespeare's London frequently had to drive the kites off—as the black vultures have constantly to be driven off from the food-stalls in the markets of Central American cities today.

The English name of the wind-borne contraption that is flown at the end of a string is another testimony to familiarity with the original kite in days gone by.

Although the red kite is no longer a city bird anywhere, it continues to be not uncommon on the Continent today. Its numbers are reduced to insignificance, however, by comparison with those of its congener, known as the black kite in Europe, as the pariah kite in southern Asia. In spring, when all but a few immature gulls leave the Lake of Geneva for their nesting grounds, their swarms are replaced by the swarms of black kites newly arrived from Africa to nest in the trees of parks along its shores. (For years, a pair nested in a tree a stone's throw from the house in which my family and I then lived.) During the following three or four months, the skies over lake and city are busy with the comings and goings of the kites. Although they don't descend into the streets of the city or land on its roofs like the gulls, they are constantly criss-crossing it over the house-tops.

Above the lake, in summer, they fly back and forth with great wing-strides, often chasing one another or playing in the air, repeatedly tumbling to the surface to pick up dead fish, of which there appears to be no shortage. One sees each bird, then, while remaining on the wing, feed by presenting the fish to its beak with one talon. Having eaten the choicest part, it lets what remains fall to the water.

While it seems to prefer the vicinity of water, it is a common bird everywhere except above tree-line in the mountains, where it occurs not at all. Suddenly, however, before the end of August, its numbers are gone, for it returns to Africa as soon as its young are well on the wing.

I have already cited the account of this kite given by Brown and Amadon on the remarkable adaptation it has made to human community in India.[1] They describe it as one of the world's most successful birds of prey, and probably the most numerous raptor of its size anywhere. No other species appears as adaptable to all sorts of conditions over a large part of the world. "In Africa and the East [it] is commensal with man to a greater or lesser degree, and it is usually far more numerous in and around towns and cities than in thinly populated country."

✦

I have, in these pages, been approaching with some delicacy what is a delicate matter when characterizing the national bird of the United States of America. The fact is that *Haliæetus leu-*

---

1. See page 49, above.

cocephalus, the bald eagle, is properly to be grouped with the kites rather than with the typical eagles (genus *Aquila*). It belongs to the *Milvinæ*, a subfamily of the *Accipitridæ* that includes the black kite. Within this subfamily, it is genetically associated, not only with the other fish-eagles of the genus *Haliæetus*, but also with the common Brahminy kite of southern Asia, *Haliastur indus*, which also has a white head. The fish-eagles and kites are part of one taxonomic group. So it is that if, like Adam, we were naming the species for the first time, we would do better to call *H. leucocephalus* a kite rather than an eagle.[2]

The most abundant bird of prey in southern India, after the black kite, the Brahminy kite, is more a bird of the rice paddies, and of the seaports, where it is a scavenger. Like the black kite, it tends to be commensal with man. Indeed, in all its ways and habits it seems intermediate between the black kite and the bald eagle, although it is the size of the former.

Let us be sensible. The American national bird, although more precisely a kite than an eagle, is still one of the most impressive representatives of bird-life on this planet. If the kites can claim it for their kind, that is not to its dishonor but to their honor.

Indeed, the kites would have reason to complain of the bad name given them in English literature. Some of them are among the most beautiful of birds, the most graceful of flyers.

Picture, now, a bird the size of a herring gull but the shape of a barn swallow. It has the pointed wings and long forked tail of the swallow. Its round head and its underparts are liter-

---

2. See B. Campbell and E. Lack, eds., *A Dictionary of Birds* (Vermillion, S.Dak.,, 1985), p. 275; and A. Landsborough Thomson, *A New Dictionary of Birds* (London, 1964), p. 356.

ally as white as snow, as are the inner parts of its wings, seen from below. The rest of it—back and tail and flight feathers—is glossy black, but with an iridescence of purple, blue, or green and a bloom like the bloom on grapes. (Note that the swallow, too, has such iridescence.) Its beauty of form and color is matched by the beauty of its flight, which is that of the swallow raised to higher dimensions. This is the swallow-tailed kite, which I have seen in forest clearings of Central America, but which also occurs, sparsely and ever more rarely, in Florida and up the coast as far as South Carolina. If anyone should propose it as a candidate for the title of most beautiful bird in the world, what other bird could be offered as a rival?

The species of kite are wonderfully varied, many are second only to the swallow-tail in beauty, and most are not scavengers at all.

The most buoyant and gracious of the raptorial species in the Swiss Alps, short of the golden eagle, is the so-called honey buzzard, which is a misnamed species of kite. It lives in the spruce forest, where it feeds largely on the larvae of bees and wasps, which it digs out of their nests with its feet. Every summer, where we live in the Alps, we watch its display flight, rhythmical like music. It dives, rises, and at the top of its pitch repeatedly claps its long wings together over its back—then repeats the performance.

Even the large red kite, with its long forked tail, is lovely on the wing, and is not primarily a feeder on carrion. Shakespeare must have known it only under the circumstances in which it became, for a long time, a garbage-collector in city streets.

✦

Although the golden eagle may pass high over a city that lies, perhaps, beneath its notice, it will never enter a city. Unlike other birds that take refuge in cities, the European population is not threatened in its native habitat; for the Alps, like other great mountain-ranges, are not permanently inhabited by our kind above tree-line. It is true that high bowls in the mountains have been violated by all the machinery that goes with the development of ski-slopes, that inns have been planted here and there as way-stations for mountain-climbers, and that the climbers themselves have multiplied. But so much of the mountains extends, still, beyond and above these stains of civilization! Viewed from the peaks, as the eagle views them, the grand landscapes still show the littleness of man.

Moreover, since it confines its hunting to areas above the level of human agriculture, the golden eagle has escaped decimation by the pesticides with which our kind has been extirpating other raptors over wide areas of the earth.

The bald eagle, by contrast, was brought close to extermination everywhere except in Alaska by the use of the pesticides in North America after World War II. Accumulating especially in the fish that formed so large a part of its diet, they almost brought its ability to reproduce to a complete halt. Restrictions on the use of pesticides have since reduced this threat, so that the great bird is beginning again to produce and raise young. But now the increasing density of human settlement makes it increasingly hard for it to find places where it can nest undisturbed. It has not, as yet, gone so far in its adaptation to human civilization that it could nest in a city park— although, if it survives, the day might come when it does.

Indeed, it has already shown itself to be something of a city bird. In the times before its virtual extirpation in the United States, it used to be seen off Manhattan in winter, floating down the Hudson River on cakes of ice. One of its centers of abundance was along the Potomac River from Washington south—but including Washington. Day after day it used to visit the Tidal Basin, close to the center of the city, alarming the many wild ducks that roosted on it, putting them to flight by its passes low over the surface. At National Airport, one might see up to half a dozen perched in trees at its edge. Once I looked out of my office window to see a full adult, with white head and tail, flying down Virginia Avenue, not far above the level of the house-tops, apparently on its way to pay a call at the White House.

In Europe, the white stork nests on chimneys in the center of bustling villages and towns—or on special platforms that our kind has erected for the reception of its nests. Might not the national bird be similarly accommodated, some day, in the national capital? A pioneering pair, following the precedent of the chimney swift, the pileated woodpecker, and the goosander, might be the progenitors of a new line of city eagles. That noble raptor, the peregrine falcon, already nests on skyscrapers in the midst of some of our largest cities, on both sides of the Atlantic, taking advantage of special accommodations made for it.

I once saw the nest of that great fish-hawk, the osprey, like a stork's nest on the chimney of a house at Montauk Point, Long Island. And I have seen the snowy owl, a bird not far below the size of the bald eagle, roosting day after day on a ledge of a building in the heart of the city of Washington.

If the American eagle is to survive the spread of our civilization, it may have to make an adjustment that other great fowl, including the Brahminy kite, have showed themselves capable of making. But it will need our good will.

✧

One basis of hope that the American eagle might adapt to civilization is not only in its relationship to the Brahminy kite, but also in the fact that two of its own congeners have already done so. The African fish eagle, which resembles it closely, nests in trees on the grounds of country houses in Africa, so that the householders can look out of their windows into the nests. And the white-tailed eagle of the Old World is to be seen among the ships and the machinery of Scandinavian harbors.

May we not foresee, then, a day when the American eagle, at home on Washington's Mall or in its riverside parks, builds its nest on a wagon-wheel or similar structure that the public authorities have erected atop a high pole for the purpose? There is a precedent in the structures erected to accommodate the nesting of storks in European towns and villages.

✧

The largest of the birds that used to inhabit the Alps, with a wingspread of nine feet (compared to seven for the golden eagle), is the lammergeier, which is thought to be related to the vultures and therefore bears the alternative name of "bearded vulture." It is being raised in captivity for release in its former haunts, and we shall now see whether it can re-establish itself in today's Alps.

I have seen the lammergeier, in its immensity, soaring over the shrine of Delphi, and could find it in myself to regard it as a sacred bird.

In 1981, a free-flying pair of griffon vultures, at the Salzburg Zoo in the Austrian Alps, nested four or five miles away, reared two young, and returned with them to the zoo to feed on its offerings.[3] This suggests the possibility of a permanent commensal population. How splendid it would be to have this majestic species on the wing above our towns and cities—as in Africa, as in the Guadarrama Mountains of Spain, as in the mountains of Greece!

✧

Evolution, as represented by the Tree of Life, is basically *divergent*. In more superficial terms, there is also the phenomenon of *convergent* evolution. Although sharks and porpoises belong to altogether different limbs on the Tree, being respectively fish and mammal, by convergent evolution they have developed roughly the same body-shape and lateral fins to cope with the same liquid environment.

So the diving petrels of the Antarctic Ocean and the alcids (auks, murres, puffins, etc.) of the Arctic seas are alike in looks and behavior, although quite unrelated.

A notable example of convergence is that of the old World vultures, belonging to the family that includes the hawks and eagles (*Accipitridæ*), and the New World vultures, which belong to an entirely different family (*Cathartidæ*). Although on separate branches of the Tree of Life, both generally have, by convergent evolution, the unfeathered heads that are adapted to poking their beaks deep into the greasy insides of carcasses.

The convergence between Old World and New World vul-

---

3. *British Birds* 74 (June 1981): 260.

tures goes beyond this. The hunt for carrion requires hours of soaring over a large territory in search of dead animals, great or small. So it is that the Old World vultures and the vultures of the New World are alike adapted, beyond all other birds, for soaring flight, and so share the shape of body and wing best adapted to such flight. Both groups tend to be sociable in their hunting, spreading their numbers across the sky so that, when any one of them discovers a carcass, the others, noting its descent upon it, can follow. Consequently, the spectacle of soaring vultures widely scattered across the sky, or all heading downward toward one point, is essentially the same in Africa and South America, although the birds of the one area are so far separated, genetically, from those of the other.

To the innocent eye, then, the great condor of the Andes is closely akin to the great soaring vultures of Africa and southern Asia, although so unrelated genetically.

⁕ VII ⁕

The identification of birds is complicated by the fact that they wear different costumes in different circumstances. It is only a minority of species that, like the crows or the chickadees, have but one plumage which both sexes wear from first fledging. Many, like the American robin or the mallard, have an immature plumage that is replaced by the adult plumage before the end of its first year. Some, like the golden eagle or the bald eagle, will acquire the adult plumage progressively, in a series of annual stages covering, in their cases, four or five years. The wandering albatross may take thirty or more years to acquire, year by year, what one must call the plumage of its old age; but this is exceptional, if not unique.

Among many species one has to distinguish between a plumage for the breeding season and a winter plumage—as in the cases of loons and grebes, of sandpipers, and of the American wood warblers.

There are rules that govern these distinctions. One is that, in species vulnerable to predation, when both parents alternate in incubating the eggs or brooding the young at an open nest, they wear the same plumage, plumage that makes them inconspicuous to predators. When only the female attends the eggs and young, the female plumage is inconspicuous while the breeding plumage of the male may be the opposite—as in the case of the scarlet tanager, the mallard, or the peacock.

Another rule is that, when male and female plumages differ,

immature plumage, in both sexes, corresponds to that of the female rather than that of the male.

Still another is that related species tend to resemble one another more in their immature than in their adult plumages, so that we take for granted the greater difficulty of identifying the immature. One has only to look through the plates of any field-guide to see that this is so. It is striking, for example, in the plates showing the hawks, the gulls, the American wood warblers, the thrushes, the finches and sparrows. (In some of these cases, the representations of female or winter plumages have to serve as surrogates for unrepresented immature plumages, to which they tend to correspond.)

Take, more specifically, the plates showing adult and immature hawks of the genus *Buteo* or *Accipiter*. In both cases, the immature have an almost uniform vertical brown streaking on breast and belly, where the adults have a variety of different patterns or colors. Or look at the thrushes of the genus *Turdus* (or almost any other thrush genus). The immature plumage of all shows spotting underneath, even in the cases of the American robin and the European blackbird, which are not spotted but evenly colored in the adult plumages.

What is the explanation?

In the development of the individual from the egg-cell to full adulthood, it goes through stages representative of earlier stages in the evolution of its kind. Thus the developing egg-cell that became the reader of these lines passed through a stage in which it had some general characteristics of its piscine ancestors, later lost. (At one stage it had what appeared to be gill slits, such as fish require for underwater breathing.) Accordingly, immature birds generally represent an earlier stage in the evolution of their respective species. Since evolution is a process of divergence, the earlier stages are less differentiated than the latest, and it is these earlier stages that the immature plumages represent. We may be sure that our present *Buteos* and our *Accipiters* are respectively descended from a common ancestor that was streaked below, that the present species of *Turdus* are descended from a common ancestor that was spotted below. The common ancestor, in each case, is represented by the immature plumages of the present species, which therefore resemble one another as the more developed, and consequently divergent, adult plumages do not.[1]

✧

I now come to a point that scientists are bound to disregard because, while it has a certain self-evidence, it is hardly susceptible of testing or proof—and, indeed, it does not belong to the domain of science at all.

The point is that, where immature and adult plumages differ, the adult plumages are the more beautiful. Again, I hardly have to argue this because it is self-evident in the plates of any field-guide—at least if it represents artistic fidelity and competence. (Many don't.) Look at the gulls, look at the diurnal birds of prey, look at the pheasants, look at the herons, look at the birds-of-paradise—look at any group at all. Surely the whole of mankind, asked for judgment on the widest array of specific cases, would agree. (In Chinese and Japanese painting,

---

1. Some of us were brought up on "Hæckel's Law," that "ontogeny recapitulates phylogeny," that the history of the individual recapitulates the history of its kind. This is too crude if one takes it to be true in detail, because ontogeny itself is constantly modified in the course of evolution, so that it is only general characteristics of the sub-adult stages of one's ancestors that manifest themselves in ontogeny. Nevertheless, it represents the most fundamental kind of insight, by which our understanding is enhanced.

There is also Karl Ernst von Baer's observation, over a century and a half ago, that, the earlier the stage of embryological development, the more do distant classes resemble one another.

we see a preference for the depiction of birds in the adult plumage. I have not tested Hottentots or Andaman Islanders, but their taste for bright colors and long plumes suggests that they, too, would agree.)

See, now, what the immense implications of this are for the direction of evolution, given that the adult plumages represent a more advanced stage than the immature plumages. The conclusion I come to is that evolution is a process of perfection along its various branches.

I have, at this point, taken beauty as the test of perfection, while recognizing the difficulty of defining it in other than subjective terms, in terms of what appeals to me and my fellow humans. But the difficulty is not insuperable, for the appeal is not arbitrary. It is based on the elements of pattern, symmetry, rhythm.[2] By objective tests, the swallow represents a higher order than its ancestor, the tongue worm—and this, we may well believe, is the basis of its greater aesthetic appeal.

Compare the earliest bird known, the archæopteryx of 135 million years ago, with the golden eagle or the swallow of our own day. Awkwardly constructed as it was, by comparison with modern birds, its utmost achievement in flight would have been to flap, with great effort, from one tree to its neighbor. The fossil record is all we need to confirm the statement that evolution is a process of perfection along its various branches. This process is still going on, which explains why birds are more beautiful in their adult than in their immature plumages.

I have no doubt that the nightingale sings more beautifully than did its ancestors.

---

2. This is the theme of my *Out of Chaos* (Boston, 1977); but see especially pt. 6, chap. 1.

To suppose that the eye with all its inimitable contrivances for adjusting the focus to different distances, for admitting different amounts of light, and for the correction of spherical and chromatic aberration could have been formed by natural selection, seems, I freely confess, absurd in the highest degree. . . . Reason tells me, that if numerous gradations from a simple and imperfect eye to one complex and perfect can be shown to exist, each grade being useful to its possessor, as is certainly the case; if further, the eye ever varies and the variations be inherited, as is likewise certainly the case; and if such variations should be useful to any animal under changing conditions of life, then the difficulty of believing that a perfect and complex eye could be formed by natural selection, though insuperable by our imagination, should not be considered as subversive of the theory.

—CHARLES DARWIN[3]

Of our five senses, sight is the most highly developed among us human mammals. Yet it is not nearly as highly developed among us as among some or almost all birds.

In days gone by, I kept a tame red-shouldered hawk, named Akbar, that sat on my gloved fist when I went out walking in the country. Often, during the seasons of hawk migration, as soon as I took him out of his shed he would abruptly fix his gaze on some point in the sky where I could see nothing. Training my binoculars in the direction of his gaze, then, I would often discover the tiniest speck, invisible to my naked eye, representing a migrating hawk which had caught his attention so quickly that it must have been conspicuous to him.

On some occasions, although I could find nothing even

---

3. *The Origin of Species*, 6th ed. (New York & London, 1915), chap. 6, pp. 223–24.

through my binoculars, he might take several minutes to follow what was unseen by me from horizon to horizon, thus showing that he was keeping his eyes on what was presumably a migrating hawk that took so long to cross the sky because it was so far away.

Each of our own human eyes has a depression called the fovea, in the center of the retina, where the image received by the eye is most distinct. The cells called cones, which register bright light, are concentrated at the fovea. Hawks, however, have two foveæ per eye, and a far greater concentration of cones in each fovea, the common European buzzard or buteo having five times as many as we have. Moreover, they have larger eyes for their size, those of the buzzard being about the same size as our own, although it is only about a fiftieth of our weight. The hawk I see as no more than a flake against the sky may, I daresay, be seen in the very detail of its feathers by the hawk on my hand.

I quote the following from an earlier account I gave of Akbar:

At the point where the path comes out of the woods into the open, Akbar suddenly shied like a frightened pony, leaping from my hand as if a paper-bag had been exploded in his ear. At the time I thought nothing of it, attributing it to some mental vagary in his over-alert consciousness, but when, on subsequent walks, the same behavior was repeated at the same point, I began to take notice. At first I could find no external explanation of his baiting, but finally one day I lighted on the cause, and subsequent observations confirmed my discovery. On a hillside nearly a quarter of a mile from where the path emerged from the woods a herd of cows belonging to a neighboring farmer was usually grazing. Akbar had never before seen cows, and those great horned beasts that seemed so far away to me that I had not even noticed them were close enough to startle him by their proximity.[4]

When I see a hawk take flight at my approach, even though I am still a long distance from it, I remind myself that, to it, I am more prominent than it is to me.

The eyes of the hawk order are undoubtedly much better than our own at estimating distance (the two foveæ of each contributing to this) and at adjusting focus. This is important for the raptor that stoops upon its prey at a speed which may attain two hundred miles an hour, keeping it in focus all the time.

The evolution of the eagle's eye, from a light-sensitive nerve in a primitive ancestor, is an example of evolution as a process of perfection.

✧

For the most part, mammals and birds can be divided between predators and prey, the pursuers and the pursued. The pursuers have both eyes in front and pointing forward, as in us humans, while the pursued have their eyes on the sides of their heads, repectively pointing in opposite directions. Dogs and cats, hawks and owls, see straight ahead, in the direction of

---

4. *Birds against Men* (New York, 1938), p. 64.

pursuit, while mice and rabbits, pigeons and parrots, see in a wide circle—forward, to each side, backward, above, and below—so as to see a pursuer from whatever angle he may come.

We know from our own experience how the two eyes of the predators are co-ordinated to give one image, forward looking as we are. The two eyes see what lies ahead from slightly different angles, so that they present two slightly different images, but the two images are combined in the brain to make one stereoscopic image that, in consequence of the difference, shows it in three dimensions. Such binocular vision would be hard for us to imagine if we did not experience it ourselves. What is harder for us to imagine, if not impossible, is how the world appears to a parrot, which, because its two eyes look in opposite directions, sees two different scenes with little overlap. Dr. Pettingill (on what basis I don't know) says that, when birds so equipped "cock their heads to examine an object on the ground," they look "with one eye while vision in their other eye is suppressed." The vision of the other eye is suppressed, not by shutting it, but presumably by a sort of inattention of the brain, which is altogether intent on the image that the first eye presents. But this applies to a particular situation. Dr. Pettingill also says that such birds can "view binocularly by peering straight ahead at an object with both eyes." The American woodcock, which has its oppositely directed eyes far toward the back of the head, may have better binocular vision to the rear than forward.[5] It could better tell where it has been than where it is going.

---

5. Olin Sewall Pettingill, Jr., *Ornithology in Laboratory and Field,* 4th ed. (Minneapolis, 1979), p. 128; and A. Landsborough Thomson, ed., *A Dictionary of Birds* (London, 1964), p. 864.

One supposes that all birds must have binocular vision in the forward direction so that, in flight, they can see where they are going. I note, however, that the zigzag trajectory characteristic of the woodcock would enable it to see in the general direction of its flight with first one eye and then the other. Some or perhaps all species of snipe, which also have their eyes farther back than other birds, also have the habit of zigzag flight.

✦

Although the life of a parrot is such that it does not need the eyesight of an eagle, and although it sees with maximum clarity only out of one eye at a time, its monocular vision is better than our own binocular, good as that is. In Westchester County, New York, in days gone by, I used to go on walks with a yellow-naped parrot called Lorenzo, whom I carried on my hand. He served me as a pointer, pointing migrating hawks as Akbar did. An account of this that I wrote at the time has the following:

> Lorenzo sat upright on my hand, attentive but at ease, observing nothing and seeing everything. It was enough for me to watch him . . . . Suddenly his head would be cocked on one side, one yellow eye looking up into the transparent sky. My gaze would immediately follow his, straining until it found its object, one lone speck circling or floating down the wind; or sometimes a whole cluster of specks like water-bugs against the blue. Then Lorenzo would be moved to my shoulder and, like the mere man I was, I would raise my glasses to search for the characteristics by which I might differentiate genus and species, age and sex.[6]

---

6. *Birds against Men* (New York, 1938), p. 190.

The owls, some of which have eyes larger than our own in bodies so much smaller, also have excellent vision, with special adaptations for seeing in the dimmest light. They are even more remarkable, however, for a hearing so acute, and so sensitive to direction, that at least one species, the barn owl, can catch a mouse in total darkness, locating it exactly by the faintest rustle or squeak.

Studies made of barn owls in a laboratory have revealed the bases of its competence. It can tell the direction of a sound in the horizontal plane by any difference in the time it takes to reach its two ears respectively, and any difference in volume— or, of course, by the absence of such differences. For the vertical direction of the origin of a sound, as opposed to the horizontal, it depends on a difference between the way the two ears, and the channels leading to them, are pointed, a difference whereby the right ear is better at receiving sounds from above, the left ear at receiving sounds from below. Thus a difference in the loudness of a sound reaching the two ears respectively, adjusted for its horizontal direction, gives the owl its vertical direction.[7] One has to conclude that, in most circumstances, the information provided by the ears of a flying owl, in pursuit of a moving mouse, is equivalent to what the eyes might provide in daylight. It has a positional picture of the mouse provided by its ears alone.

The facial ruffs so characteristic of owls are simply feathered reflectors of sound, serving to direct high-frequency sounds, especially, into the aural openings. They are the equivalent of a cupped hand behind one's own ear. The more noc-

---

7. The circumstances are more complicated in their totality and detail than in my simplified account. See Eric I. Knudsen, "The Hearing of the Barn Owl," *Scientific American* 245, no. 6, pp. 83ff.

turnal species of owl (e.g., the barn owl) tend to have more developed ruffs than those that also hunt by daylight (e.g., the snowy owl).

One genus of hawks, the harriers (represented by the northern harrier in North America), depend in considerable degree on hearing to capture their prey, since they hunt over grass-covered ground on which rodents are hidden from sight but might be heard. In addition to having especially large ear-openings, they also have facial ruffs.

Hearing, as well as vision, represents the perfecting process of evolution.

✦ VIII ✦

There are more things in heaven and earth, Horatio,
Than are dreamt of in your philosophy.

In the breeding season, when the male scarlet tanager as-
sumes his bright plumage, the female continues to wear the
green that makes her so inconspicuous in the leafy woods. We
explain this by saying that, since he never has occasion to visit
eggs or young, the scarlet that makes him conspicuous to predators endangers him only; but she, having to sit on an open
nest, or to come and go in connection with the feeding of the
young, might betray the nest to predators if she, too, were conspicuous. This tends to be confirmed by the fact that, among
vulnerable species in which both partners attend to eggs and
young, both are inconspicuous.

While this offers a satisfactory explanation of why the
female does not wear scarlet, it leaves the question of why the
male does. Presumably he wears it, even though it reduces his
chances of survival, for some practical reason. The scarlet
must confer some advantage—and confer it in the breeding
season, since in other seasons the male is as inconspicuously
colored as the female.

What advantage does it confer?

Darwin explained this sort of thing as representing sexual
selection. The female is attracted to those males that are the
most impressive in their appearance—impressive to her, and

incidentally to us as well. The competition of the males for her acceptance, based on such impressiveness, is made obvious in the competitive displays of peacocks before peahens. Presumably it is the same with the tanagers. The competition of the males is in brightness of color. The females have, over the generations, chosen the most brightly colored males as mates, and such selection has led to ever greater brightness in males.

However, the answer that females prefer bright males only begs the question by raising it in other terms. If the male's scarlet, in spite of the disadvantage to which it subjects him with respect to predators, conveys a practical advantage by making him more attractive to the female, how is it that the female has developed a preference that goes counter to selection for survivability? After all, the female's preference is itself the product of the natural selection that is directed at success in the competitive struggle for survival. Why, then, does the female prefer males that, because of their conspicuousness, have less chance of survival over males, that, being less conspicuous, have more chance? Here natural selection, manifested at one remove in sexual selection, seems to favor a disadvantage in the struggle for survival.

I have referred to the tanager as if it were a special case, but what it represents is the rule over the widest range of species, including ducks, cranes, pheasants and grouse, birds of paradise, hummingbirds, and a large proportion of all passerine  species. It appears to be particularly the rule for many species that inhabit tropical forests. The question I have raised also applies to the European kingfisher, and any number of tropical kingfishers, in which both sexes, rather than one alone, are made vulnerable by spectacularly bright coloration. In fact, it appears to be the birds that least need to fear predation—hawks and eagles, owls, gulls, and petrels—that tend to wear drab plumage.

Consider the peacocks, which engage in competitive displays before the peahens, raising in a fan the enormous trains, with their pattern of ocelli, that appear to be products, not of natural selection as we know it, but of deliberate design and spectacular artifice. Such trains, so far from representing any practical utility, except as they impress the peahen, must have absorbed considerable energy in their growth, and can hardly be anything less than heavy impediments when the cock has to take off suddenly to escape a predator. Using metaphorical language, one would say that Nature was impractical when she designed the peacock. One would say that the design came to prevail in spite of the natural selection that is directed at survival.

There is a possible answer to the question we confront here. It is that the males who can mount the most impressive displays are also the healthiest and strongest, consequently the most fit to be perpetuated by the female's choice. According to this argument, the most brightly colored tanagers would also be the healthiest and the strongest. I daresay this is true. But why should health and strength be manifested in such disadvantageous forms when they could be manifested in less disadvantageous forms, as they are in any number of successful species, from the hawks that engage in display flights to the male house sparrows that engage in competitive aggressiveness toward the females? Why does the male tanager have to be so conspicuous when the tufted titmouse, an equally successful species, is not?

Another argument is that, in such direct competition

among males as leads to fighting, bright colors serve the purpose of intimidation. But most species show that intimidation can be achieved as well by less costly means. In any case, peacocks do not direct their displays at one another. Moreover, their displays cannot be said to intimidate potential predators. A peacock does not display before a leopard that might otherwise pounce on it.

The question I raise here applies as well to the birds that, like the nightingale and the mockingbird, attract the attention of predators by their elaborate and continuous singing.

In sum, I know of no answer to the question within the strict bounds of the theory of natural selection.

✧

Just as man can give beauty, according to his standard of taste, to his male poultry, or more strictly can modify the beauty originally acquired by the parent species, can give to the Sebright bantam a new and elegant plumage, an erect and peculiar carriage—so it appears that female birds in a state of nature, have by a long selection of the more attractive males, added to their beauty or other attractive qualities. No doubt this implies powers of discrimination and taste on the part of the female. —CHARLES DARWIN[1]

It is a feature of human psychology that we tend not to ask any question until an answer is in sight—and, if someone nevertheless does ask a question to which no answer can be suggested, we are impelled to dismiss it with some offhand response improvised on the spur of the moment. We are impelled to show that the question either does not arise, or presents no

---

1. *The Descent of Man* (New York, 1915), p. 214.

problem, or is not worth considering. (Not all the professionals, in any field, like to admit that there are things they do not know.) But the question I raise here was raised implicitly by Darwin in Part 2 of *The Descent of Man*, and he also suggested, implicitly, the direction in which an answer might be found. While I cannot answer it, neither can I dismiss it. Let me therefore set forth its principal implication.

In the above discussion I have assumed that birds see color much as we do, that the scarlet of the male tanager appears to the bird-hunting hawk and to the female tanager much as it does to us. This has been confirmed by experiments on diurnal birds—to the extent that it can be confirmed at all. (The lack of bright colors in nocturnal birds, by contrast with diurnal ones, reflects the fact that their vision, like our own night vision, generally excludes color.)

We are on more difficult ground when we ask whether birds share our æsthetic sense, for we hardly know how to define that sense in ourselves. Most examples of sexual selection suggest that they do—that, just as they see the colors we do, so they respond as we do to what we regard as æsthetically satisfying. It would be hard on any other grounds to explain the responsiveness of the peahen to the display of the cock—and this example could be multiplied a thousandfold by examples from most of the taxonomic groupings.

Again, the song of the male nightingale has a musical sophistication that the female must appreciate as we do, since it is designed for her and for rival males, rather than for us. In her own terms, she is moved by it as we are in ours.

The best known manifestation of what appears to be an æsthetic sense in birds, corresponding to our own, is provided by various species of male bowerbirds, which construct on the ground, not for our appreciation but for that of the females, elaborate bowers, which they then proceed to decorate. Some bowers entail the construction, by means of interwoven twigs, of great architectural structures with floor and walls. Inside such structures, brightly colored objects—berries, flowers, the wing-cases of beetles—are ranged as in an art exhibition, the flowers being replaced by fresh ones as they wilt. Some bowerbirds even engage in the art of painting. They make the paint by mixing vegetable pigments with earth and saliva, then apply it to the walls of their bowers by means of improvised paint-brushes of dried grass held in the bill. Do they not, in this, show an æsthetic sense such as we are able to recognize because we know it in ourselves?

We cannot put ourselves inside the heads of birds to see the world as they do, but we can draw some conclusions from their physiology, and others from their behavior. Thus the  structure of a bird's ears, together with the bird's reaction to a sound, tells us that it can hear. The structure of the eye, together with the bird's responses to particular colors, tells us that it sees colors. The pains that the male bowerbird takes to create a work of abstract art, apparently for the moving effect it has on the female, suggests that both sexes have an æsthetic sense akin to our own.

With one minor exception, said Darwin, "if we look to the birds of the world, it appears that their beauty has been much increased since that period, of which their immature plumage gives us a partial record."[2] Sexual selection has contributed to

---

2. *The Descent of Man* (New York, 1915), p. 499.

the increase in beauty referred to by Darwin. It has contributed to the perfecting process of evolution.

✧

The brain of a bird is very different from the human brain, and in many respects clearly inferior. In some respects, however, it may be superior, just as the eyesight of many birds is superior to our own. There is good evidence that the brains of many birds respond to external contingencies far more rapidly than our own.

What Einstein introduced into physics, the relativity of time, appears to have its counterpart in biology. Einstein imagined a space-traveler who, from the point of view of an observer on Earth, lived according to a time-scale that was retarded in proportion to his velocity. We have reason to suppose that vertebrate species live according to different time-scales, even though we attribute the differences to the innate reaction-time of each rather than to the mechanics of the cosmos. It is important, for what follows, to note that, for an organism living according to a relatively accelerated time-scale, all movement appears to be slower, as all movement is bound to appear faster to those who live according to a relatively retarded scale. Where two minutes of another species correspond to one of our own, in its consciousness, a development that occupies a minute of our time will seem to it to take twice as long, therefore to be twice as slow.

It seems evident that any passerine bird is in the opposite situation from that of the space-traveler with his relatively retarded time. For it, time passes at, let us say, five times the rate it does for us humans, and its life-span is perhaps one-fifth of ours by our own reckoning. An event that takes one second according to us might, on the bird's time-scale, take what we would regard as five seconds; while, conversely, the dozen

years it might live would, in terms of the rate at which time passes for it, be the equivalent of sixty for us. On its time-scale it is as long-lived as we are on ours. (This means that the bird would see everything in what was slow motion from our point of view. For it, the Earth might take a hundred and twenty hours, rather than twenty-four, to complete one revolution.)

How can I, who have never inhabited the body of a bird, come to such a conclusion? The answer is that I come to it by plausible surmise based on observation of avian behavior.

Let me begin by an observation accessible to us all. Many cages for small birds have swinging perches suspended from the ceilings on short wires, so that, when a bird suddenly takes off from one, the perch is made to oscillate violently. However, even when it is swinging with such rapidity that the human eye can hardly follow it, one may see a flying bird land on it with perfect ease and precision, as if it had been motionless. The explanation is that the bird sees everything in relatively slow motion. In its perception, the swing is moving at, say, one-fifth the speed that it is moving for us—and it perceives itself, too, as flying at one-fifth the speed we attribute to it. A very slow-motion movie would, I surmise, reproduce what the bird itself experiences. As bird and perch approach each other slowly, in this movie, the bird extends its feet and, opening its toes, closes them again when the unhurried contact has been made.

What appears as a prodigious feat on our time-scale poses no problem on the bird's.

A dramatic demonstration of the like phenomenon occurs whenever a goshawk, hunting its avian prey, shoots through dense woods in a long, jagged trajectory like a stroke of lightning. What seems a wonder is the rapidity with which it must have to take account of every successive obstacle in its path, twig or branch, and so avoid them all by constant swerving. Again, I find no other explanation than a shortness of the bird's time-scale, by comparison with ours, that entails the experience of a slow motion altogether different from the swift motion our own observation registers.

We understand many birds better if we understand that everything passes in slow motion for them, so that a minute of their lives is to them as several minutes of ours are to us. But I have no way of surmising as confidently that this is true of all birds as it is of a linnet or a goshawk. I know of nothing in the behavior of the wandering albatross, for example, that would justify a like conclusion. Living as it does on food that it finds floating on the surface of the sea, it never has occasion to put on such a performance of precision at high speed as the linnet in the cage or the goshawk in the woods, and its life-span approaches if it does not actually equal our own. Neither does the basic relationship among all members of the class *Aves* give reason to attribute accelerated time to all, since we have cause to suppose that the members of our own class, the mammals, differ widely in this respect. There is reason to believe, for example, that a shrew lives its short life at high speed—and consequently in slow motion—by contrast with a sloth, although shrew and sloth are both mammals.

The giant tortoises of the Galápagos Islands, which move with excessive slowness and have life-spans of a century and a half, may live, in their consciousness as in their movements, on an exceedingly retarded time-scale that makes the passage of time faster for them than for us.

This is as far as I can carry the matter without actually putting myself inside the skin of finch or hawk, of shrew or sloth or tortoise.

✦ IX ✦

Between three and four thousand million years ago, in some pool of warm water, a molecule too small to be seen, even under an optical microscope, showed the first sign of life in its ability to be fruitful and multiply, reproducing itself by growing and splitting. This was the beginning of the process of evolution, which, exceedingly slow at first, has been accelerating ever since.

A hundred and fifty million years ago, after ninety-six percent of the time from that beginning to our present, the first organism that could be identified as a bird appeared in the form of a rather reptilian creature with feathers, to which we have given the generic name *Archæopteryx*. If not absolutely the first creature that could be called a bird, it was the first of which we have evidence on the fossil record. Presumably its ancestor had been one of those scaly reptiles that, running on its hind legs only, had a long tail to balance it. In the course of its evolution, certain scales on its forelegs and tail had grown longer, looser, and more flexible, and had become shredded—until they were feathers that helped sustain it in such leaps as it made through the air, perhaps from tree to tree or from tree to ground.

Upon the appearance of the first self-replicating molecule, what ghostly observer could have foreseen the archæopteryx? Upon the appearance of the archæopteryx, what ghostly observer could have foreseen the Alpine swift, the golden eagle, or the hummingbird? From its single beginning, the tree of life has branched and rebranched in innumerable directions, all representing the accelerating process of perfection that still goes on. The æronautical mastery of the birds that followed the archæopteryx has already gone so far toward perfection that one wonders what remains to be accomplished. So the ghostly observer of the archæopteryx, seeing the power of levitation attained by a leaping lizard, seeing how its leaps went to lengths that no one could have foreseen, might have wondered what remained to be accomplished. He could hardly have anticipated the way of the eagle in the air, of the sandpiper over the sea, or of the nighthawk in the night sky over the cities of mammals that did not yet exist. As we now know, however, the archæopteryx was not an end but a beginning.

Is the eagle an end? Or the sandpiper? Or the nighthawk? Is our own species an end?

There is, by definition, a limit to how far the process of perfection can go along any particular branch. Perhaps the hummingbird—that specialized descendant of the archæopteryx, direct or collateral—represents a dead end in the achieved perfection of its kind. There are, it is true, many species of hummingbird, the twig-ends of the branch, with various adaptations of a minor nature that correspond to various environmental circumstances; but it is hard to imagine any important further progress along the main line they represent. Surely the possibilities for the improvement of their flying ability must have narrowed, by now, to very little.

Our own kind is much further from being finished because it is not so specialized. Moreover, the mental abilities that distinguish our kind remain, still, so inadequate to the comprehension and mastery of this vast realm of being that they seem remote, as yet, from the realization of the possibilities they represent. In a curious way, the most advanced of animals is the most imperfect, far more distant from the perfection of its kind than the hummingbird is from the perfection of its more limited kind. So it is that the lizard is closer to its own perfection than we are to ours. (This gives point to D. H. Lawrence's remark that, if a man were as much a man as a lizard is a lizard, he would be more worth looking at.)[1]

In what follows, I am concerned with features of evolutionary adaptation in the birds that represent approaches to the perfection which is, by definition, a dead end.

✧

The house sparrow, which has become so wholly adapted to life within the precincts of our civilization, shares our imperfection. It is hard to believe that this or that hummingbird could become more beautiful than it already is, or that it could make further progress as a flyer within the limits of its mode of flight. The realization of its possibilities has, in its measure, entailed the foreclosure of its possibilities. But one cannot say of the less specialized house sparrow that no room for improvement remains either in its beauty or in its flying ability. Its line, like our own, is still far from a dead end.

The house sparrow belongs to the African family of weaverbirds. Its ancestors, perhaps in savanna or woodland, wove a variety of covered grass nests in the branches of trees. When our own kind took to agriculture, the sparrow took to the human and agricultural environment that was thereby artificially

---

1. Here, as elsewhere, I use the term "perfection" to refer to a directional process toward what is a notional end only.

provided. And when our kind took to the building of great cities, the sparrow accompanied it into those cities, eventually abandoning the natural wilderness entirely and, in its evolution, becoming adapted exclusively to the human association. No modern hummingbird would have been sufficiently plastic, because unspecialized, to show a like adaptability.

Imagine, now, a close relative of the house sparrow, resembling it basically, that becomes specialized for life at high altitudes, above the level of normal human habitation. In order to cope with the cold at such altitudes, it becomes somewhat larger than its weaverbird ancestors and its cousin, the house sparrow—since, the greater its size, the less the surface, in proportion to volume, through which its body heat radiates away. It does not, however, become so large that it can no longer lead the basic life of a sparrow, having a sparrow's mode of flight, feeding like a sparrow, as would be the case if it became the size of an eagle. Presumably, for it evolution in this direction has been cut off.

Like other birds that have become adapted to life in snow and ice, this sparrow acquires a partially white plumage—in its case, a patch of white on each wing and white outer-tail feathers that show chiefly when wings and tail open in flight.

Its most specialized adaptation, however, is the development of larger wings and a larger tail, so as to be able to support itself more easily on the thinner air of high altitudes.

I have now described the snowfinch, which occurs above tree-line in all the great mountain-ranges from the Pyrenees to the Himalayas. It is the only passerine bird in the Alps, and the only bird of any sort except the ptarmigan, that never descends below tree-line, even in winter. Indeed, it may in winter be driven to seek its food at higher altitudes, where the peaks, scoured by the wind, are too steep to hold snow.

I have only once seen the snowfinch at its nest. A pair was nesting in the gutter of the stone building that houses the Swiss customs at the Great St. Bernard Pass, just as a pair of house sparrows might have done in precisely the same place if the altitude had been less. Around them was the bustle of humanity, with the lines of cars passing into Switzerland or into Italy.

The snowfinch prefers to nest in the walls or under the eaves of such huts or sheds as our kind builds above tree-line, generally for the accommodation of cattle or the making of cheese in summer. Otherwise it resorts to crevices among the rocks.

Except when nesting, the snowfinches associate in flocks of twenty or fifty or more. On the Alpine scree they are well camouflaged, resembling in their color and markings so many stones among the rocks and stones. Because of their large wings, they are buoyant in flight as the house sparrow is not, rising in the thin air like bubbles through water.

Unlike the sparrows, they walk and run, rather than merely hopping. Perhaps this represents adaptation to an uneven terrain.

It was only last August, as I write, that a friend and I climbed from the Val Ferret to the Col de Fenêtre at 2700 meters (8,858 feet). We had passed three frozen lakes and had just crossed a snow corniche to arrive at the col. All around us, growing alongside the melting snow or sometimes through it, were little beds of those Alpine jewels, the *Androsace alpina*, cushion flowers which share the snowfinch's habit of living high up amid the fields of perpetual snow. The two of us sat down to rest now, our backs against a stone frontier-marker, he in Switzerland and I in Italy. An icy wind was blowing across the col from the Italian side. I had just commented on how pleasant it was to be in a Mediterranean country again, when a lone snowfinch landed between the two of us and began walking about our extended feet, so that I thought he might hop up on them. This was the only time I had seen a snowfinch by itself. After a minute, seeing that we were not picnicking or offering it food, it flew off. Like the Alpine chough, about which I shall have something to say, the snowfinch has learned to have expectations of mountain-climbing picnickers.

The snowfinch, which so enlivens the high mountains of the Old World, long ago entered on a course of specialization that now limits its possibilities. Today, one supposes, it could no longer live and compete with other species, as its ancestors had, at altitudes below tree-line. Its wings would be too large for efficient flight in the denser atmosphere. In the rarefied atmosphere of what has become its natural habitat, however, it is free of the intense competition among many species that occurs lower down, and of the vicissitudes of human civilization. Its present habitat has hardly changed in ten thousand years, and seems the least likely to change. (Even the depths of the oceans have suffered marked ecological changes from pollution.) It has found a niche in which its future is limited but relatively secure.

Do we not appreciate the house sparrow the more for knowing the snowfinch, and the snowfinch the more for knowing the house sparrow? Let us constantly broaden our knowledge.

✦

The Alpine accentor, in the course of its evolution, has also become specialized, but to a lesser extent, for the islands of grass, the escarpments, and the scree slopes of the high Alps. A bird that might be taken for a sparrow, if it did not have the thin bill of a thrush or warbler, it also shows the buoyancy in flight associated with what the æronautical engineers call low wing-loading. On almost any climb above tree-line in summer, making one's way along the steep slopes that overlook the bowls and valleys below, one comes upon a pair or family group of these stout and stalwart birds. The size of the house sparrow, they are more rotund and more delicately marked in various tones of gray with chestnut streaking on their flanks. Because they are quite unspectacular, one appreciates them inadequately on first acquaintance. But any bird life in such an immense emptiness comes to have its appeal by association. When the climber pauses high up, overlooking what seems to approach all space and all eternity below, and sits on some rock to enjoy a snack of bread and cheese, the approach of the accentor in search of remnants, shuffling about over the rocks, is like a visitation from Olympus. It shows little fear of us humans, but will not, as a rule, stay in one place long enough for extended observation.

In winter, the accentor comes down a thousand meters to the huddled Alpine villages of the lower slopes, where it makes itself at home on the balconies of chalets, becoming commensal with our kind.

This one species occurs in all the high mountains of the Old World, from the Pyrenees to Fujiyama. I have no doubt that it has a commensal association with the llamas at Lhasa in Tibet, not distinguishing Buddhists from Christians.

Its only congener in Europe is a mousy little bird that skulks in lowland hedges and shrubbery, seeming always fearful, coming into the open on top of hedge or bush only to contribute to its environment the thin jingle of its song. The traditional English name for it is hedge sparrow, but, since it is not a sparrow, its other vernacular name, that of dunnock, is to be preferred. Habitat is a character of species no less than song or color, and here the difference between the two accentors is somewhat like the difference between a being that is at home in the kingdom of the sky and one that lives fearfully at the bottom of a well. (I speak as the agoraphiliac I am.)

✦

Among the birds that have become specialized for the rarefied air of the heights, none is as buoyant as the Alpine chough. Its multitudinous companies drift among the snow-clad peaks from the Atlas Mountains of Africa and the Pyrenees to the Himalayas. Swarms of choughs may be seen swimming through the sky like shoals of fish that turn first this way, then that. Indeed, they appear more like fish that move in a medium heavier than themselves than like birds that have to support themselves in a lighter medium. It is as easy for them to rise as to sink, the latter requiring them to fold wings and tail. This is because they have such extensive wing and tail surfaces. Nevertheless, when one does suddenly flap, as a fish may suddenly work its fins in a dart, the action of its wings is rapid and deep. In this they are like their corvine relatives, the jackdaws.

Anyone who sees the Alpine chough for the first time recognizes it as a member of the crow family by the black plumage in a bird that, if not as large as our familiar crows, is still of another order of magnitude than either an American or a European blackbird. But it has touches of color in its bright yellow bill and its red feet. When one has learned to appreciate the princeliness of its appearance and behavior, and when one hears its high, musical cheeping or trilling, it comes to seem quite different from those commoners, the European carrion crow and the American common crow. It constitutes a class of its own in a family that also includes magpies, jays, and nutcrackers.

One may see several hundred in a swarm that drifts across an azure sky or along the contours of grass-clad slopes, detachments separating themselves from it here or there, merging into it again, all cheeping and trilling as if the sky were full of tiny guitarists. A detachment will dive down through the atmosphere, turning this way and that, wings half folding and unfolding, to inspect a dome of grass that crowns the crags, some landing for a moment. Then, finding nothing, all let themselves be carried up through the atmosphere again, perhaps half a mile. It is as if the invisible currents of air were harnessed to their bidding.

Sometimes the flocks disappear into the clouds, and then one understands the utility of the constant peeps and creaks, which enable them to keep together. (High notes carry better than deep ones, as the yodelers of the Swiss Alps know.)

When, in our climbing, we reach a summit of our destination, and sit down to rest, perhaps breaking out food and drink, we can expect that several choughs will join us, touching down on nearby rocks, striding or jumping about, cheeping all the time. They expect from picnickers, if not handouts, orange-peels, apple-cores, and crumbs left behind.

Indeed, the choughs of the Alps, whatever may be the case elsewhere, are basically commensal with our kind for a large part of the year. Every morning, large companies descend for many miles through valleys and gorges to towns far below—for example, the towns of the upper Rhône Valley—where they swarm over the housetops, landing on ledges and balconies, looking for spillage or handouts. In mid-afternoon, allowing enough time before dark, their flocks rise circling and float up again through the valleys and across the passes to roost overnight in crevices of the highest peaks. They live a life of daily transhumance, Sundays included.

The Alpine chough, attracted by the prospect of such scraps as may be found in the wake of our humanity, has followed mountaineers up to 8,100 meters (26,575 feet) in the Himalayas, an altitude at which men need oxygen masks, but choughs don't.

(Neither do the bar-headed geese that have been recorded in flight over the Himalayas at nine thousand meters, about 29,500 feet, which is slightly higher than Mt. Everest. Anyone who sees them in flight at the lower altitudes of their winter quarters, in Nepal and India, has occasion to be impressed by the size of their wings and the associated buoyancy of their flight—just as in the case of the snowfinches.)

In the Pennine Alps, and elsewhere as well, the choughs have until recently lived in part on the garbage-dumps at the high mountain inns. (So have those other corvids of our mountains, the ravens.) In the past decade, however, provision has been made to carry the garbage down by cable-transport into the lowlands for processing. The result is a sudden marked diminution in the numbers of choughs (and some diminution of ravens too). At places where choughs were always present in swarms, one now sees few or none.

So it is that, because the ways of our civilization change so abruptly, such commensal dependence as the choughs have developed is not without hazards.

Over a great part of the world, ravens used to be among the principal scavengers of our civilization, living on the garbage of London and other cities, towns, and villages. Even in recent times I have seen their loose companies about the housetops of Greek towns, or on the garbage-dumps at their outskirts. Now some of the garbage-dumps are gone, and with them the ravens.

I think that, when the town of Delphi, which harbors the shrine of Apollo, stopped dumping its garbage over the cliffs below, not only lammergeiers, griffon vultures, and ravens departed from it. So did the god.

Look at the way the Japanese have painted birds. . . . The system is simple. They have sat down in the countryside and have for a long time watched the birds in their flight. By dint of watching them they have ended by understanding their movements. —AUGUSTE RENOIR

The wing of a bird in flight has no fixed shape. Its shape is constantly changing in response to the constantly changing pressures of the air through which it moves, or to suit changes in course or speed. Man has never invented anything at once so strong and so delicate, anything so sensitive in its adjustment to the continuously varying conditions it has to meet.

A bird's foot, being a relatively simple mechanism, changes shape only to the extent that it folds or unfolds at set joints; but the changes of shape that a bird's wing undergoes are fluid throughout. It shortens or lengthens, narrows or broadens, and the curvature of its surfaces is constantly being modified, subtly or drastically, slowly or suddenly. No one shape is quite retained for more than a moment. This is true even of the great albatross that banks and wheels, rising into the sky and coasting down to skim the running waves on what we think of as set wings. A myriad small adjustments, at least, are constantly occurring.

---

1. This chapter, with the illustrations by the author, first appeared in the *Virginia Quarterly Review* as "The Flight of Seabirds."

Because the wing of a bird in flight is motion itself, no photograph or drawing can do it full justice. The representation can never be more than a half truth. When, for example, I watch the flying seabirds at close range on their nesting grounds in the Shetland Islands, far to the north of the Scottish mainland, it is clear that the wing of the fulmar is much narrower than that of any gull. But I have photographs of flying fulmars in which the wing appears as broad as a gull's simply because of the way it is being held at the moment.

Fulmar

✦

Scientists may speculate whether, in the course of evolution, form or function has come first. However they have developed in relation to each other, they are inseparable, each giving meaning to the other. Nothing illustrates this more beautifully than the flight of seabirds.

Birds have various modes of moving through the air, with all gradations between them. In one mode commonly associated with fast and direct flight, the wings move rapidly up and down like two sticks oscillating through a short arc. This is typified by the members of the auk family—the razorbill, the murres, the black guillemot, the puffins—and by the fulmar, which alternates a succession of rapid beats with intervals of gliding.

In another mode, the bird progresses by successive impulses that represent distinct wing-strokes separated by pauses. For this pulsating mode, which is the normal mode of such typical terns as the common and the Arctic, the wings are jackknifed. This is to say that they reach out and forward from the body to the midway joint, from which they angle sharply back—like a half-folded jackknife. The frigate or man-of-war birds of the tropics, although so much larger than the terns, also fly this way. So, for the most part, do shearwaters.

The pulsating mode, as it is less mechanical, is more flexible than the oscillating mode. The fulmar tends to move straight and fast, whereas the tern may alter course and speed with each stroke of its wings. It can stop and go, dodge and dart, pausing to see whether a glimmer in the wave below is a fish to dive at—and, if it is not, resume its full speed at one stroke. (However, when flying long distances in calm air, as when migrating, the terns have a shallower and steadier wing-beat. At other times, when simply holding their positions in the face of the wind, they wave their wings slowly and gently.)

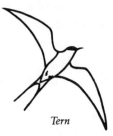

Tern

✧

Whether a wing is jackknifed or straight does not depend on the species. A fulmar can jackknife its wings like a tern in order to tip down into the updraft along the face of a cliff; a tern can extend its wings like a fulmar to brake its flight.

The jackknifed wing, however, is the only one for sprinting. This is exemplified among land birds by the peregrine falcon, among seabirds by the parasitic jæger. The jæger, living largely by the practice of piracy, combines power and elegance in its flight as does no other bird. One circles on extended wings off the high bluffs and cliffs of Shetland, where puffins are nesting. When it sees a puffin approaching from the sea with a sand eel held crosswise in its bill, it sets off suddenly with jackknifed wings striking down in stroke after stroke, seeming to reach its top speed with the first stroke. The puffin may or may not have time to save its provender by plunging vertically to the water and disappearing beneath the surface. If not, it can save itself from being disabled only by dropping the fish for the jæger to pick out of the air in one long and lovely swoop.

The reason why the wing had best be jackknifed for sprinting is clear. The outer half, because it lags after the inner on the down stroke, slopes upward from joint to tip, its long terminal feathers being further bent upward by the pressure of the air beneath them. To the extent that this upward sloping segment points backward rather than merely outward, it propels the bird forward on each downbeat. Moreover, the air is less resistant to the jackknifed wing.

✧

In seabirds of any size the oscillating mode of flight tends to go with relatively narrow wings, which may be beaten the more rapidly because, by their narrowness, they present less surface to the resistance of the air. They cut through it as broad sails could not. The limited wing-surface, in turn, may represent an adaptation to an environment of strong winds. For, the stronger the wind that supports it, the less wing-surface the bird needs, while more surface than was needed would make its progress against the wind that much slower and more difficult, as well as requiring more muscle.

Because the wings of gulls are broader than those of fulmars, their wing-beats are relatively sluggish. While they cannot, therefore, match the fulmar's speed in straightaway flight, or the sprinting ability of the

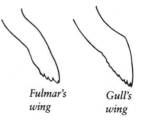

Fulmar's wing   *Gull's wing*

jæger with its jackknifed wing, they are better suited to drift-ing and soaring in light winds. The wings of the fulmar are adapted to the great winds of the ocean far from land, those of gulls to onshore and offshore breezes.

✧

Even on a windless day a gull can rise lightly from sea or land in one spring; and it can feed from the surface of the water in hovering flight. A fulmar cannot properly hover (al-though it can hang in a sufficiently powerful updraft against a cliff), and it can take off from the water only by a taxiing run in which it has to work its feet strenuously. This means that, where gulls and fulmars are feeding together on offal dumped from fishing boats, the gulls get there first.

On windless days in Shetland the fulmars launch themselves with difficulty from the stone walls on which they sometimes roost. Unable to rise until they have gained sufficient air-speed, they lose altitude until they are bumping along the ground, running with their feet and beating their wings stren-uously to achieve the speed at which they become, at last, com-pletely airborne.

It is surely no accident that the kittiwakes, which are the only gulls that range the open ocean rather than the shores, have wings distinctly narrower than those of the typical gulls. The difference in the breadth of their wings and the conse-quent mode of their flight is well seen during the nesting season in Shetland, when their presence in the same sky with typical gulls makes di-rect comparison possible.

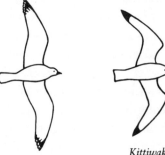

Herring Gull

Kittiwake

✧

The kittiwakes do not, however, have the fulmar's mode of flight. They are, rather, intermediate between the typical gulls and the terns, although closer to the former. They hold their wings partially jackknifed as a rule, the inner halves extending forward more than in the typical gulls, although the outer halves do not sweep back as sharply as in the terns. They have a more pulsating flight; and when, like the typical gulls, they circle on fixed wings outstretched, they do so only momentarily.

The kittiwakes are like the terns, too, in their diving ability. I have seen a flock of them over a school of fish diving vertically from on high with wings folded, all disappearing beneath the surface for an appreciable time.

In flight they differ from the typical gulls, and from the terns too (although not from fulmars), in that head and body appear to be all one, as if the birds were flying torpedoes, whereas the typical gulls and the terns show the constriction of a distinct neck separating head from body.

The pulsating flight of the kittiwake combines irregularity with regularity in the way that distinguishes everything in nature from anything that is merely mechanical. The appreciation of this as a fundamental characteristic of nature was well expressed by the French painter Renoir in his notebook—as quoted in the biography written by his son. Noting that the great artists of the best periods imitated nature in that "they created irregularity within regularity," he wrote:

> The Earth is not round. An orange is not round. None of its segments has either the same form or the same weight. Open them and they will not have the same number of seeds, and the seeds will not be alike. . . . I do not want a column to be any rounder than a tree.

I never saw a kittiwake fly in a perfectly straight line, even when purposefully maintaining a single direction. Instead, there is a slight wavering or yawing that goes with a habit of tilting from side to side. The wing-beats have a regular rhythm within which they are irregular. The tilting and wavering is accompanied by an alternate quickening and retardation of the beat, but all within the limits of the measure.

In his *Principles of Natural Knowledge*, Alfred North Whitehead wrote:

> A rhythm involves a pattern, and to that extent is always self-identical. But no rhythm can be a mere pattern; for the rhythmic quality depends equally upon the differences involved in each exhibition of the pattern. The essence of rhythm is the fusion of sameness and novelty; so that the whole never loses the essential unity of the pattern, while the parts exhibit the contrast arising from the novelty of their detail. A mere recurrence kills rhythm as surely as does a mere confusion of detail.

Anyone who studies music or poetry appreciates the need of such irregularity within regularity, but how few have traced it back to nature as Renoir did! Verse with only a mechanical beat is doggerel rather than poetry, and no one sensitive to the quality of music could long endure listening to a composition that had no more rhythmic variation than a metronome— even though the variation occurred, as it must, within a fixed measure.

✦

Usually, when we think in these terms, we talk about nature as the great artist; but nature provides the original that serves as model to the human artist. It came first. While we may say that the flight of the kittiwake is like music, with rests, grace-notes, and turns that have their place within a larger rhythmic pattern, it would better accord with the logic of being to say of a piece of music that it followed the model provided by the flight of the kittiwake.

Of two wings with the same surface area, one may be long and narrow, the other short and broad. Species that inhabit woodland, like our common crow, need the shortest possible wings to thread their way through the mazes of branch and twig; so that their wings, in order to have enough surface, must make up in breadth for what they lack in length.

I know of no reason why breadth of wing, in itself, should make possible a steeper angle of climb. It is a fact, however, that crows, pheasants, American black vultures, herons—all such broad-winged birds can climb the air at remarkably steep angles. Perhaps the explanation is that the lifting power of a wing is greater in the degree to which it is cambered (arched from front to back), and a broad wing lends itself to cambering as a narrow one does not. A heron, its wings broad and cambered alike, is a slow if buoyant flier. In the days of falconry, when a peregrine was loosed against a heron the heron would try to remain above its pursuer by climbing steeply if slowly, while the falcon, with its flat rate of climb, would try to get above it by the speed of its flight in wide circles that some-times took it beyond the horizon. (The narrowness of wing in the fulmar, the razorbill, the murres, and the puffins explains their difficulty in rising from the surface of the water at all in the absence of a powerful wind. A mallard, with its broad wing, has no difficulty.)

Two European birds, the lapwing and the wallcreeper, have wings that seem broader in their outer than in their inner halves. In the case of the lapwing this is associated, presumably, with its ability to take off from the ground at the steepest angle; in that of the wallcreeper with its habit of flying almost vertically up the face of a cliff, albeit in stages with brief touch-downs between.

The lapwing is the only bird I know with broad and rounded wings that holds them in the jackknifed position for its typical flight. In that flight it swerves constantly, from side to side or down and up, as if governed by a succession of erratic and ecstatic impulses.

(Despite my exceptional citation of the little wallcreeper, my observations apply, in general, only to birds above a certain size; for the relation of surface to volume is different, creating different circumstances, in small birds like swallows and sparrows).

Because the biggest seabirds, such as the great albatrosses, have little or no need either to fly slowly or to climb steeply, they have the longest and narrowest wings of all. For narrow

Albatross

wings are more efficient than broad ones in that the turbulence of the air caused by the passage of their tips is less. The wings of the wandering albatross are like strips of whalebone that extend up to eleven and a half feet from the tip of one to the tip of the other. A great albatross may spend most of its life sailing with outspread wings over the unbroken ocean of high southern latitudes, circumnavigating the earth repeatedly without coming to land. Under such circumstances it never has occasion to climb steeply or to reduce its speed drastically. However, when it does have to come to land for its nesting, every other year during its adult life, its inability to reduce its speed sufficiently may cause it to tumble tail-over-bill upon touching down. To become airborne again, it generally has to walk to the top of some hill where it can take off into the wind.

✧

In my experience, when a great albatross remains in attendance on a ship that has a following wind, it has to circle the ship continuously, presumably because it cannot reduce its speed enough to hang in her wake like a gull.

Many large land birds with broad wings, such as the turkey vulture, glide and soar on wings that slope upward in a flat V. It may be that this enables them to soar in tighter circles upon thermals, those columns of air that rise above ground heated by the sun. Seabirds, however, have no such need to limit their soaring to tight circles. All those that are large enough to glide or soar hold their wings more or less bowed, the tips below the horizontal. This bowed position is often extreme in the great albatrosses; so that, seeing them in their wheeling flight, one could believe that their springy whalebone wings had invisible weights hanging from the tips. The reason, I surmise, is that in the bowed position the body is supported chiefly by the inner part of the wing, which is held against the air with less muscular strain. It is somewhat as if the outer segments should be partially folded away when they were not needed. Bowing the wings shortens sail.

Jaeger

Many seabirds have an excess of wing-surface for ordinary circumstances. This is demonstrated on the nesting grounds of parasitic jægers, kittiwakes, and Arctic terns, the young of which may rise in flight, albeit with some effort, when wings and tail are still only half grown.

Even when they are not sprinting, jægers and terns may hold their wings in the jackknife position, half folded, simply because they have too much sail to carry it fully spread to the wind. They reef it so as not to be blown away.

✦

The wings of birds are so infinitely varied in form and use that one finds exceptions to almost any generalization one 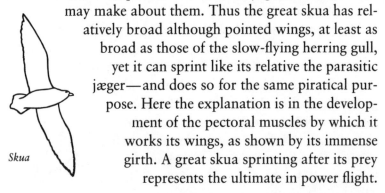may make about them. Thus the great skua has relatively broad although pointed wings, at least as broad as those of the slow-flying herring gull, yet it can sprint like its relative the parasitic jæger—and does so for the same piratical purpose. Here the explanation is in the development of the pectoral muscles by which it works its wings, as shown by its immense girth. A great skua sprinting after its prey represents the ultimate in power flight.

Skua

✦

Even the most extreme specialization comes up against limits set by circumstances. Perhaps the wings of the wandering albatross would be shorter than they are, and therefore not have to be shortened by bowing in ordinary flight, if they were not needed, however rare the occasion, to brake the great bird's momentum when it comes in for a landing on its nesting grounds.

Every bird, even the albatross, must sometimes come to rest, on water or on land, with wings folded away. In fact, if the choice is between being airborne all the time or never, it must be never. Ostriches and kiwis lost the power of flight completely at some stage of their evolution, so that they could be more efficient on land. The penguins lost it so that they could be more efficient under the water.

The penguins once had dual-purpose wings for ærial and underwater flight alike. To the extent that their wings were adapted to ærial flight they were inefficient for underwater flying, and vice-versa. Because underwater flying, by which they caught the marine organisms on which they lived, was more important in the particular environmental circumstances of their lives, they at last specialized in the former at the expense of the latter; so that today their wings are mere flippers, excellent for propulsion under the water but useless for rising above it.

One may doubt that circumstances in the Northern Hemisphere, to which they are confined, will allow the surviving members of the auk family to lose their powers of ærial flight. (The circumstances of the penguins, confined to the Southern Hemisphere and breeding where there were no mammalian land predators, have been quite different.) The great auk did lose its powers of ærial flight, like the penguins, but the result was its extermination when man at last found his way to its nesting grounds and proceeded to harvest its numbers. The last one died in 1844. (I daresay whoever killed it proceeded to wash his hands.)

The wings of all the other auk species—of razorbills, murres, guillemots, puffins, dovekies—are compromises between adaptation for ærial flight and adaptation for underwater flight. Requiring a smaller surface for the denser medium but a larger for the less dense, they undoubtedly have too much for the one and too little for the other.

Looking down from the top of the cliffs on which they breed, one may sometimes see the common murres in their underwater flight, which presumably typifies that of all the auks.

The wings are outstretched only to the wrist-joint, the outer segments that bear the primary flight feathers being folded so as to form spikes that point backward—thus reducing the entire wing surface for the denser medium but adding to the forward propulsion of each wing-stroke. The bird proceeds by separate successive strokes, each shooting it forward and being followed by a pause during which it loses momentum.

✦

If the fully spread wings are too large for use under water, they are smaller, relative to the stout bodies, than they ideally should be for use in thin air; which accounts for the rapidity with which the auks generally have to beat them in overwater flight. A black guillemot taking off from the surface of a tossing sea sets its wings a-humming—but it may still collide with successive wavecrests before it gets free, if it does at all. (When a guillemot, murre, or puffin has to get out of the path of a ship quickly, it generally does better by diving, but it is often moved to try the alternative nonetheless.)

On the other hand, skimpy wings may be excellent in a high wind, especially if it takes the form of a powerful updraft against the sea-cliffs where puffins, murres, and razorbills nest. The wings of these three species are excessively narrow; but in a roaring updraft against the cliffs they may suffice to hold the bird suspended without themselves moving, except for an occasional tremble.

Razorbilled Auk

✦

One may say of the typical gulls (as distinct from the kittiwakes) that they are not true seabirds because, roosting every night on land, they never range far out from shore. The auks and the fulmar, however, when they are not at their nesting grounds, live night and day on the open ocean, sleeping on the billows, drinking only salt water. If it were not that they had to come to land for nesting, their wings might become more specialized, adapted to the pelagic environment only. As in the case of the albatross, however, their wings must retain at least enough braking ability to enable them to land on hard ground without crashing.

Fulmars, for all that they are birds of the open ocean, are highly specialized for patrolling flight along the cliff-faces. There is a mystery here in that, although they spend a large part of their lives in such flight, one can see no useful purpose that it serves. It seems a pure pastime.

The patrol requires spectacular dexterity in flight. Recently, standing at the top of a cliff where it was broken inward to form a trap for the howling wind from the sea, I spent the best part of a quarter hour watching a fulmar ten or a dozen feet from my face. It was riding the turbulence of the updraft like a cowboy riding a bucking bronco. It kept its position, simply allowing itself to be carried backward or forward, to one side or another, within strict limits of distance—like a kite at the end of its string. Wings and tail were constantly responding by twisting and bending, folding and unfolding, to the sudden buffets from below.

Another fulmar in the same wind would slip slowly, inch by inch, down the broken, sloping face of what was half cliff, half bluff, almost in contact with the surface of which it followed every irregularity. When it had got low enough, it would allow itself to be lifted quickly to the top to resume the slow, slipping descent.

✧

One of the fulmar's special adaptations for such cliff-hanging is a capacity to make its wings tremble in order to give some slight extra lift or propulsion. The wholly unrelated puffin, which also patrols the cliffs and steep bluffs (although generally farther out), has the same adaptation. It may be able to hold its position in the most powerful updrafts without actually beating its wings; but it is so much smaller than the fulmar, with so much less mass to oppose to the wind, that it must frequently vibrate them to avoid being blown backward.

Fulmar

The puffin may also vibrate its wings when it coasts at a steep angle from its high nesting-sites to the water. The larger razorbilled auk does the same. On the other hand, razorbills also descend to the water by simply waving their wings slowly like big butterflies.

The puffins commonly nest in burrows on steeply sloping bluffs that face the sea. Unlike the birds that nest on ledges of vertical cliffs, they generally come in to the landing platforms in front of their burrows from above, and their wings are barely adequate to such an approach. They beat them hard as they draw close, but at the last instant seem to give up and drop vertically, hitting the ground with a thump.

If there were talking puffins with philosophical minds (and, more than any other birds I know, they look as if they had philosophical minds and could talk), I can imagine one saying: "We have to use our wings to catch fish under water; we have to use them to take off from the water and fly over it; we have to use them, in accordance with the rules of our nature, to sail in the updraft against cliffs and bluffs; and we have to use them to come in to our landings at our nesting burrows. Because our wings could not possibly be perfectly adapted to all these purposes at once, they necessarily represent a compromise. We are not, after all, like the penguins, who have to do only one thing with their wings—and *can* do only one thing with them."

✧

Although there is less mystery about the flight of birds than there used to be, the wonder continues everlasting. Aerodynamic theory, as developed in connection with the design of aircraft, together with slow-motion pictures of birds in flight, has taught us much about how the wings of birds lift and propel them. (To the mind of one layman, at least, what remains truly mysterious is the flight of a butterfly opposing its weight to the wind.)

The everlasting wonder is partly in the structure of the wing itself, at once so light and so strong, so complex in the multitude and diversity of its parts, so flexible in their mutual connections and articulations, so harmonious in their combination to form a single whole. The wonder is also in the diversity of its adaptations to the diverse requirements of so many species, from the storm petrel, hardly bigger than a butterfly, to the great albatross.

What is no less a wonder is the control exercised by the bird, from some center in its head, over all the shifting surfaces at once. The constantly changing shape of a bird's wing in flight, as of its tail, is produced by a complex and subtle interaction between the varying pressures of the air and the impulses that come from inside the bird itself in the form of what is equivalent to countless decisions per instant.

The sensitivity of all these infinitely rapid adjustments may be seen in the flight of a wandering albatross over the waves of the South Pacific. Banked on set wings, the bird moves perpetually in a great wheel tilted toward the surface, so that it rises high only to come sweeping down again until the lower wing-tip is skimming the waves, almost touching them but never quite. The point of the wing follows the moving contour of a sloping and rippled wave-surface so closely that it actually describes the successive ripples, rising and descending over them in a fluttering, shimmering movement. Although moving at high speed and remaining virtually in contact with the rough and changing contour of a running wave, the wing-tip never touches it, never scores the surface, never gets hooked in it. The bird's eye, which in any case must be occupied elsewhere, is some five feet away from the wing-tip. Surely the delicate control involved in this constantly repeated performance must be in automatic response to the changing pressures of air between the wing-tip and the surface of the sea.

✧

All the species of albatross that practice this mode of flight, and the giant petrel as well, may be seen doing the same. It is almost as wonderful to see shearwaters or prions moving swiftly through the avenues between the waves of a tossing sea, remaining below the crests, threading the troughs, skimming the moving slopes—wings bending, dipping, flickering, changing in all their 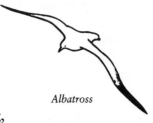 contours and dimensions—yet not once touching the wave surfaces that toss all around them.

Albatross

I have referred to the wings of birds as if they alone were involved in the performance of flying. But the tail and, in many

 species, the webbed feet are involved; and, indeed, the whole body of the bird is engaged in the intricate and subtle process.

✧

The delicacy and beauty of bird-flight is the product of long evolution. The quasi-reptilian ancestors of the present birds must have been ludicrous in the long hops they made with the aid of their primitive wings and tails. But evolution is a process of progressive perfection, along different paths for different categories of life. Regarding it as such, we must conclude that these birds have come closer to the ultimate perfection of their being than we have to the ultimate perfection of our own.

Murre

In the flight of a bird is a whole philosophy, if only we could read it right.

APPENDIX

The inquiry that follows is based on the thesis that we know less than we pretend to know about how the life I have been describing evolved.

In Hans Christian Andersen's tale, "The Emperor's New Clothes," a couple of fraudulent tailors, arriving at the emperor's court, undertake to make an outfit of beautiful new clothes for him which will have one peculiar property, that of being invisible to anyone who is unfit for his job. They proceed to go through the motions of weaving thread on what is, in fact, an empty loom. Then they go through the motions of arraying the emperor in what they refer to as the beautiful new clothes they have made, inviting his admiration and that of his entire court. Despite the fact that no one, including the emperor, can see any clothes at all—since there are none—all vie with one another in their praise of what each pretends to see, believing that all the others see it, for each is concerned not to reveal to the others that he is unfit for his job.

One who is himself a professor may be allowed to observe that this tale illuminates the situation in which most of us professors find ourselves. The students who sit at our feet assume that, as professors, we know the answers to all the questions that might be asked about the respective subjects we profess. It follows that, if any professor should show himself at a loss for an answer, he would, in the eyes of the students, be showing himself unfit for his job. So each is impelled to answer every question with an air of authority, whether he knows the answer or not. Ordinarily, there is nothing consciously fraudulent in this behavior, because, after years of practicing it, a professor falls victim to his own pretense. Truth tends to become, for him, whatever he hears himself saying—or, at least, whatever he and his colleagues all say.

No doubt most of my readers, at this point, will tell themselves that I must be exaggerating. I have not, however, said that we professors are ignorant of the answers to all the questions that arise. Most of us do hold in our heads a vast amount of simple factual information, such as the composition of sodium chloride, or the date of Napoleon's death, or the nesting behavior of the peregrine falcon. It is on the fundamental questions of why things are as they are that we suffer from the ignorance that we are driven to cover by the pretense that becomes so habitual as to be unrecognized even by ourselves.

I don't confine this to professors. It tends to apply to all those who make their careers as professional experts on matters of which they can have only limited or uncertain knowledge.

The classic statement of this ignorance is in Plato's *Apology*, his account of the defense made by Socrates at his trial, where Socrates reports his experience in making the rounds of the supposed authorities in various fields to interrogate them. In each case, he was surprised to discover in what degree their apparent knowledgeability was a pretense that they, themselves, had come to believe. He concluded, almost incredulously, that his own unique distinction resided in the knowledge of his own ignorance.

Plato reported this twenty-five centuries ago, but to no avail. The vainglory he revealed has not been diminished since by his public revelation of it. I suppose it does not occur to us professors that what he revealed about the sophists might apply to us as well.

✧

The pretense to fullness and certainty of knowledge has its collective as well as its individual manifestations. The professors in any field of learning form a self-promoting academic community that cultivates the distinction between insiders and outsiders. Such a community manifests the behavior of a priesthood that is moved to impress the laity with its supposed possession of special knowledge. Only those who have undergone a course of formal indoctrination in its orthodoxy over a period of years, and have at last been awarded a degree that signifies acceptance into the ranks of the order, are qualified to discourse on the mysteries over which it is alert to maintain an exclusive jurisdiction. Any outsider who shows himself outspokenly curious about some question in the field that remains basically unresolved is likely to be informed by the insiders that the answer is not accessible to the understanding of outsiders. Instead of receiving an answer to his question—where, in fact, there may be none—he will have been warned against trespass.

✧

In Darwin's day, this kind of organization of the academic world was not nearly as developed as it is today. An amateur biologist who had been trained for the ministry, and therefore largely an outsider, he came to believe in the theory of evolution by natural selection, but not without experiencing difficulty and entertaining doubts.

To my own mind, the fundamental truth of the theory is unquestionable. Given what we now know about genetics, given the fact that accidents to the genotype cause changes that manifest themselves as mutations in the phenotype, natural selection is bound to occur. A mutation favorable to the survival and self-perpetuation of the phenotype will tend to become established in place of the less favorable genetic arrangement that had preceded it. So evolution has, in fact, progressed by the cumulative perpetuation of the genetic changes that have caused favorable mutations, and their replacement of the old genetic arrangements over the thousands of millions of years since organic replication began. To doubt this, given the premises, would be to doubt that the universe is logical.

Accepting the rule of logic, and accepting the theory of natural selection, I am nevertheless troubled by certain manifestations of evolution that cannot easily be explained in terms of that theory. This leads me to wonder whether the undoubted process of natural selection is sufficient, by itself, to explain the whole of evolution from the original cell to the array of species that constitutes our biosphere today.

The theory of natural selection implies that every distinctive feature of every species, as it is today, came into being because it conferred an advantage in terms of the survival of the species. But we are all of us defeated, time and again, when we try to say what advantage this or that particular feature could have conferred. Indeed, we are often driven to conclude that particular features, presumably the product of natural selection, are disadvantageous. Let me exemplify.

Some toucans are endowed by nature with cumbersome bills almost as big as the rest of their bodies. One of our ablest and most eminent ornithologists, Dr. Oliver L. Austin, Jr., writes that these bills may be "as much a hindrance as a help in feeding." (They may be as much a hindrance as a help in other respects as well.) He then goes on to say that "perhaps the toucan's bill has no particular adaptive function, but developed more or less fortuitously, and its owners have been able to use it well enough to survive and prosper."[1]

See, now, what the implications of this are. Dr. Austin implies that what must have been a long series of mutations, producing ever larger bills over many generations of toucan ancestors, spread among them in spite of the fact that these mutations put their recipients at an increasing disadvantage, in the competition for survival, relative to those that, not inheriting the mutations, retained the smaller bills. Nevertheless, over a period of time the lines of the birds with smaller bills died out, being replaced by the relatively handicapped lines of those that had larger bills. The fortuity invoked by Dr. Austin might apply to one couple with its progeny—but not to untold couples over untold generations. Would not the development he surmises, on its scale, be altogether contrary to the logic of natural selection, a logic that would otherwise have to be regarded as inescapable?

Yet—there it is, the toucan's bill, to be explained; and the heavy train of the peacock, referred to in Chapter 8, above; and the train of the male quetzal, which must make it conspicuous to predators and impede its flight; and the fantastic plumed appendages of the birds of paradise, which appear to be costly impediments; and the crown of the crowned cranes, and all the other pennants, crests, and ornamental plumes worn by birds that seem to represent, rather than adaptation for survival, the ostentation of beauty for its own sake. I daresay the plumed crest of the California quail is a luxury it can well afford, but it does seem to respond to æsthetic rather than directly practical considerations, suggesting natural selection for show rather than use.

---

1. *Birds of the World* (New York, 1961), pp. 188–89.

As I mentioned in Chapter 8, Darwin explained such features as representing sexual selection, usually by the female. What this suggests is a tendency for such selection to go counter to natural selection for survivability. Although there is evidence that selection for survivability tends to prevail where a sufficiently marked conflict between the two occurs, it is philosophically interesting that there should be any selection at all on the basis of what we must regard as purely æsthetic criteria.

(I note parenthetically, for whatever relevance it may have, that extravagant coloration and ornamentation appear to be more characteristic of those taxononomic groups that have developed in the tropics, and especially in tropical forests, than of groups that have developed in open habitats of temperate or sub-Arctic zones. Perhaps this is merely an impression stemming from the fact that there are so many more species in a tropical forest, among which those that are conspicuously beautiful may constitute only a small percentage. But where does one find even a small percentage of scarlet and yellow seabirds?)

My main point is that, although we accept the theory of natural selection for survival, we find ourselves at a loss when we try to account for particular features of particular species by reference to it. Look at almost any bird—for convenience, if one wishes, in the plates of a field-guide. Then try to explain its particular features exclusively in terms of natural selection for survival. Why does this species have a black spot under each eye, or that one a chestnut crown? Why does the jackdaw have a white eye, while the eyes of other crows are black? Perhaps, if we could see a total re-enactment of the evolution of each species, we could account for the black spot, the chestnut crown, or the white eye. I note merely that the utility for survival of many features is not evident.

So one is repeatedly given occasion to wonder whether there may be some criterion of selection additional to that of adaptation for survival, a criterion making for the development of an ever more elaborate order that characterizes the evolution of life over the span of billions of years.[2] Perhaps, other things not being unequal, selection for some other objective than that of mere survival comes into play.

Now take a standard example about which there is no dispute.

None of us doubt that the facial discs of owls and harriers, which enhance the hearing on which these particular birds so particularly depend, by reflecting and channeling sound waves into their ears—no one doubts that these discs are the product of natural selection, operating independently but by convergence in the owls and harriers respectively, since the two groups are quite unrelated. But try to trace the process in imagination.

For a first try—may we suppose that, by a mutation in some ancestral owl (as in an ancestral harrier), one or several of its facial feathers came in slightly askew, in a position or positions to reflect sound into the ear, thereby providing the advantage of an enhanced hearing that caused the mutation to become permanent, replacing the feather-arrangement that had preceded it? May we then suppose that other mutations caused other facial feathers to come in askew in the same sense and to the same effect—and so on, separately in owl and harrier, until an entire facial disc had become established? Or

2. I have traced this development in part 2 of *Out of Chaos* (Boston, 1977).

may we suppose that by a single mutation, once in the owls and once in the harriers, an entire facial disc appeared in some individual, and, by the advantage it conferred, spread through succeeding generations until all owls (ditto harriers) had it?

One gropes in one's mind for a more plausible scenario, without finding it.

The inadequate plausibility of what I have described, entailing other examples, troubled Darwin, as witness my quotation from him in the form of the epigraph to the third section of Chapter 7, above, in which he said of the thesis that the eye was brought into being by natural selection, that it seemed to him absurd, although he went on to justify his belief, nevertheless, that it was true. The strain on his credulity was the greater in that he thought of the period since the evolution of life began in terms of tens of thousands of years, rather than the billions of years we now measure. The eye had had a much longer time to develop than he then knew, and length of time would also be offered to explain the development of what is so much simpler, the facial discs of owls and harriers, which may have had eighty or a hundred million years to evolve by natural selection. There are, however, developments that are hard to imagine over any finite span of time at all. (It is unimaginable, for example, that a monkey, idly tossing the letters of the alphabet into the air and letting them fall at random, would thereby ever produce the sequence of letters that constitutes Tolstoy's *War and Peace*. Anyone who doubts this should engage in the simple exercise of calculating the odds. Even if all the letters except the last had fallen right, the odds would still be twenty-five to one against it doing so. For the last two letters, the odds would be twenty-five squared, or 825 to one— and so on exponentially.) Consequently, I feel the need of a more plausible scenario for the way the facial discs were produced than I have offered above. And, although I don't doubt that the peacock's train was produced by natural selection, I feel the need to develop or supplement our present concept of such selection so as to eliminate or reconcile the contradiction with which this appears to confront us.

This much said, the reader should nevertheless be warned not to take literally the above comparison of natural selection with the possibility of writing *War and Peace* by an immense sequence of extraordinary coincidences. The sequence of coincidences that produced either of us need not have been that extraordinary, since it need not have entailed a letter-by-letter correspondence to have produced such viable organisms as ourselves. However, even when we allow for this, the apparent role of coincidence remains impressive.

In the nineteenth century, Newton's theory of gravitation was generally accepted in spite of some observations that could not be reconciled with it. One such observation was of what appeared to be an aberration in the orbit of Mercury around the sun. Another was the result given by the Michelson-Morley experiments, according to which the speed of light, as measured by any observer, was independent of his movement relative to its source. Among the cosmologists of the day, these were motes "to trouble the mind's eye"; but the general disposition was to minimize them, to suppose that somehow they would be resolved without any disturbance to the body of theory that all scientifically minded persons accepted (Newtonianism, as it might have been Darwinism). Then, however, Einstein came along with his Theory of Relativity, which explained the disturbing observations on the orbit of Mercury and on the speed of light. Einstein's theory did not invalidate Newton's so much as it built upon it and extended it.

I leave the perceptive reader to apply this experience to the difficulties referred to above in reconciling certain observations with Darwin's theory of natural selection.

BIRD SPECIES MENTIONED IN THE TEXT

accentor, Alpine	*Prunella collaris*
albatross, wandering	*Diomedea exulans*
auk, little (= dovekie)	*Alle alle*
babbler, jungle	*Turdoides striatus*
blackbird (Old World)	*Turdus merula*
blackbird, red-winged	*Agelaius phœniceus*
buzzard	*Buteo buteo*
buzzard, honey	*Pernis apivoris*
canvasback	*Aythya valisineria*
chough, Alpine	*Pyrrhocorax graculus*
condor, Andean	*Vultur gryphus*
coot (Old World)	*Fulica atra*
cotton teal (= pygmy goose)	*Nettapus coromandelianus*
crow, American	*Corvus brachyrhyncos*
crow, carrion	*Corvus corone*
crow, house	*Corvus splendens*
dipper (New World)	*Cinclus mexicanus*
dipper (Old World)	*Cinclus cinclus*
dovekie (= little auk)	*Alle alle*
duck, tufted	*Aythya fuligula*
dunnock (= hedge sparrow)	*Prunella modularis*
eagle, African fish	*Haliæetus vocifer*
eagle, bald	*Haliæetus leucocephalus*
eagle, Bonelli's	*Hierætus fasciatus*
eagle, booted	*Hierætus pennatus*
eagle, golden	*Aquila chrysætos*

eagle, short-toed	*Circætus gallious*
eagle, white-tailed	*Haliæetus albicilla*
egret, little	*Egretta garzetta*
egret, snowy	*Egretta thula*
falcon, peregrine	*Falco peregrinus*
frigate bird (= man-o'-war bird)	*Fregatta* (species)
fulmar, northern (= arctic)	*Fulmarus glacialis*
gannet, northern	*Sula bassana*
goldeneye	*Bucephela clangula*
goosander (= common merganser)	*Mergus merganser*
goose, bar-headed	*Anser indicus*
goose, barnacle	*Branta leucopsis*
goose, Canada	*Branta canadensis*
goose, greylag	*Anser anser*
goose, magpie	*Anseranas semipalmata*
goose, pygmy (= cotton teal)	*Nettapus coromandelianus*
goose, Ross's	*Chen rossii*
goose, snow	*Chen cærulescens*
goose, spur-winged	*Plectropterus gambensis*
grebe, great-crested	*Podiceps cristatus*
grebe, little	*Tachybaptus ruficollis*
guillemot, black	*Cepphus grylle*
gull, black-headed	*Larus ridibundus*
gull, herring	*Larus argentatus*
gull, lesser black-backed	*Larus fuscus*
harrier, northern	*Circus cyaneus*
hedge sparrow (= dunnock)	*Prunella modularis*
hawk, red-shouldered	*Buteo lineatus*
jackdaw	*Corvus monedula*
jæger, parasitic (= skua, arctic)	*Stercorarius parasiticus*
junglefowl	*Gallus gallus*
kingfisher, white-breasted	*Halcyon smirnensis*
kite, American swallow-tailed	*Elanoides forficatus*
kite, black (= kite, pariah = kite, dark)	*Milvus migrans*

kite, brahminy	*Haliastur indus*
kite, dark (= kite, black = kite, pariah)	*Milvus migrans*
kite, Mississippi	*Ictinia mississippi*
kite, pariah (= kite, black = kite, dark)	*Milvus migrans*
kite, red	*Milvus milvus*
kittiwake, black-legged	*Rissa tridactyla*
kiwi	*Apteryx* (species)
lapwing	*Vanellus vanellus*
lammergeier (= vulture, bearded)	*Gypætus barbatus*
mallard	*Anas platyrhyncos*
merganser, common (= goosander)	*Mergus merganser*
mockingbird, northern	*Mimus polyglottos*
murre	*Uria* (species)
mynah, common	*Acridotheres tristis*
nighthawk, common	*Chordeiles minor*
nightingale, common	*Luscinia megarhynchos*
ostrich	*Struthio camelus*
owl, common barn-owl	*Tyto alba*
owl, snowy	*Nyctea scandiaca*
peafowl	*Pavo cristatus*
petrel, blue	*Halobæna cærulea*
petrel, giant	*Macronectes giganteus*
pochard	*Aythya ferina*
prions	*Pachyptila* (plural species)
puffin	*Fratercula arctica*
quail, California	*Callipepla californica*
quetzal	*Pharomacrus mocino*
razorbill (= razor-billed auk)	*Alca torda*
robin, American	*Turdus migratorius*
scaup	*Aythya marila*
siskin (= pine siskin)	*Carduelis pinus* (= C. *spinus*)
skua, arctic (= jæger, parasitic)	*Stercorarus parasiticus*
snowfinch	*Montifringilla nivalis*

sparrow, house	*Passer domesticus*
stork, white	*Ciconia ciconia*
swallow (= barn swallow)	*Hirundo rustica*
swallow, bank	*Riparia riparia*
swan, Bewick's (= swan, tundra = swan, whistling)	*Cygnus columbianus*
swan, mute	*Cygnus olor*
swan, trumpeter	*Cygnus buccinator*
swan, tundra (= swan, Bewick's = swan, whistling)	*Cygnus columbianus*
swan, whistling (= swan, Bewick's = swan, tundra)	*Cygnus columbianus*
swan, whooper	*Cygnus cygnus*
swift ("common")	*Apus apus*
swift, Alpine	*Apus melba*
swift, chimney	*Chætura pelagica*
tanager, scarlet	*Piranga olivacea*
teal, cotton (= goose, pygmy)	*Nettapus coromelianus*
tern, arctic	*Sterna paradisæa*
tern, common	*Sterna hirundo*
vulture, bearded (= lammergeier)	*Gypætus barbatus*
vulture, black (New World)	*Coragyps atratus*
vulture, Egyptian	*Neophron percnopterus*
vulture, griffon	*Gyps fulvus*
vulture, king (Old World)	*Sarcogyps calvus (= Torgos calvus)*
vulture, long-billed	*Gyps indicus*
vulture, turkey	*Cathartes aura*
vulture, white-backed	*Gyps bengalensis*
wallcreeper	*Tichodroma muraria*
wigeon	*Anas penelope*
woodcock, American	*Scolopax minor*
woodpecker, pileated	*Dryocopus pileatus*

Other Books by Louis J. Halle

Transcaribbean

Birds against Men

River of Ruins

Spring in Washington

On Facing the World

Civilization and Foreign Policy

Choice for Survival

Dream and Reality

Men and Nations

Sedge

The Society of Man

The Cold War as History

The Storm Petrel and the Owl of Athena

The Ideological Imagination

The Sea and the Ice: A Naturalist in Antarctica

Out of Chaos

The Search for an Eternal Norm: As Represented by Three Classics

The Elements of International Strategy: A Primer for the Nuclear Age

The United States Acquires the Philippines

History, Philosophy, and Foreign Relations

About the Author

Louis J. Halle's career has spanned many years of service in the U.S. Department of State, followed by successive professorships at the University of Virginia and the Graduate Institute of International Studies in Geneva, Switzerland. He has published books in history, philosophy, and international politics. Mr. Halle is the recipient of the John Burroughs Medal (for *Birds against Men*) and the Paul Bartsch Award from the Audubon Naturalist Society for outstanding contributions to the field of natural history. He has long been a resident of Geneva.

Designed by Martha Farlow

Composed by Brushwood Graphics, Inc., in Sabon

Printed by The Maple Press Company on 55-lb. S. D. Warren's Antique
Cream and bound in Holliston's Aqualite and Kingston Natural with
Multicolor Antique end sheets